THE PIT BROW WOMEN
OF THE WIGAN COALFIELD

THE PIT BROW WOMEN
OF THE WIGAN COALFIELD

ALAN DAVIES

Frontispiece: *A series of postcards depicting surface scenes were produced in 1905 by Fletcher Burrows & Co., owners of Atherton Collieries. Here the surface browman proudly poses with his team of women coal pickers at Howe Bridge Colliery. Imagine the humour and banter he must have suffered from the women during a typical shift!*

First published in 2006 by Tempus Publishing

Reprinted in 2009 by
The History Press
The Mill, Brimscombe Port,
Stroud, Gloucestershire, GL5 2QG
www.thehistorypress.co.uk

Reprinted 2012, 2013

© Alan Davies, 2006

The right of Alan Davies to be identified as the Author of this work has been asserted in accordance with the Copyrights, Designs and Patents Act 1988.

All rights reserved. No part of this book may be reprinted or reproduced or utilised in any form or by any electronic, mechanical or other means, now known or hereafter invented, including photocopying and recording, or in any information storage or retrieval system, without the permission in writing from the Publishers.

British Library Cataloguing in Publication Data.
A catalogue record for this book is available from the British Library.

ISBN 978 0 7524 3912 9

Typesetting and origination by
Tempus Publishing.
Printed in Great Britain.

Contents

	Acknowledgements	6
	Introduction	7
one	The Early Days	9
two	The Pit Brow Lass Emerges	25
three	Protests, Deputations and Fame	41
four	Decline and Fall of the Pit Brow Lass	83
	Appendices	111
	Bibliography and Sources	125

Acknowledgements

Thanks to: Ken Wood of Heaton, Bolton, for extracts from his *Coalpits of Chowbent*; Dave Lane of Swinton for images from his pit brow women postcard collection and extracts from his *Pit Brow Lasses Scrapbook*; Mrs Teresa Redford (*née* Wall); Agnes Carter; Mrs Haselden, Leigh; Mr G. Wall; the Vaughan family; Mr Naylor; the Dick Sutton collection; the Heaton Family collection; Sam O'Neil; Mr T. Leyland, Ince; John Taylor.

Thanks also to: Wigan Leisure and Culture Trust local studies librarians Chris Watts and Tony Ashcroft; also to Elizabeth Fairclough, Diane Teskey, Philip Butler and Yvonne Webb from WLCT.

Thanks to Jonathan Smith, archivist at Trinity College, Cambridge University, for the use of extracts from the A.J. Munby diary.

Many individuals enthused about the idea of this publication and helped in various ways, including Amy Rigg at Tempus Publishing; Carol Woodward; Alf Molyneux; the late Fred Dibnah; Mark and Chris Wright; Dave Turner; Mathew Walker; Dave Duckworth; David Devine; Andrea Faulkner; Alan and Ruth Faulkner; Cliff Lovett; Mark Conroy; Gordon Sinclair; Helen Sinclair; the late William Davies; Gail Kazan; Denise Stewart; Donna Coleman; Maria McKeon; Sue Pemberton; Mr Naylor; Mrs A. Rowden; and Mr Leach; and finally, last but not least, Teresa Redford (*née* Wall) of Bamfurlong, aged eighty-five, a former pit brow woman who, just as the book was ready for the publishers, contacted me and allowed me to photograph her, copy her photographs and have a chat about her memories. Today she still has the lively character I had always expected pit brow women would have.

Introduction

Anyone who, like myself, has worked in mining will remember the unique atmosphere that the pit had and the fact that mining as a trade often passed through generations in one family. With this came a respect for those who had worked at the colliery previously, both on the surface and below. Lancashire people have always been very proud of their distinctive character and that character could be found in abundance in the mining communities. The pit brow women carried out their hard and dirty work with a smile on their faces, and their attitude is something we in Lancashire are justly very proud of.

Mining for all types of minerals over the centuries has tended to be considered a purely male preserve, yet this is certainly not the case. We know from tomb illustrations and clay tablet accounts that thousands of years ago female ore bearers were employed in the ancient copper and gold mines of Egypt, Iraq and South America. In Germany, female surface mineworkers appear regularly in paintings from the late fifteenth century onwards.

There is no reason to doubt that from the emergence of organised and documented coal-mining activity in Britain over 800 years ago, women and girls were to be found at work alongside men and boys. However, there is not a consistent growth in the numbers of women employed in British coal mines between 1750 and 1850, when coal production doubled from slightly less than 5 million tons per annum to 10 million tons. Reasons for this varied from the emergence during the Industrial Revolution of new industries such as textiles to technical changes in the mining industry itself. One example was the introduction in certain coalfields of horse-drawn rail-mounted haulage systems, thus replacing the mining woman or girls' traditional role of dragging baskets of coal along roadways.

Due to the infinitely varying nature of colliery infrastructure and methods of working over the centuries, it is not possible to speak in national terms of changes in the work demanded of mining women. While women may have been replaced by pit ponies, tubs and rails at one colliery, the neighbouring pit might not have been able or bothered to invest in such improvements and would carry on for generations using their tried and tested ways of working.

The employment of women at mines had in fact only been common in four coalfields: West Lancashire, Yorkshire, East Scotland and South Wales. In 1840 the Children's Employment Commission turned its attentions to the mining industry and began its investigative work, which was to change forever the role played by females in the industry. Their findings were published in May 1842. For the first time in a Parliamentary report graphic illustrations were employed, depicting the terrible working conditions and tasks expected of women and girls. The outrage of the general public at the revelations within the report meant that no one could really argue for the continuation of many of the workplace practices nor could they condone the presence in the mines of the lowest levels of moral behaviour. Even so, opposition to the proposal to remove women and young girls from the mines was forthcoming in the House of Lords, where many coal owners sat.

Their opposition could not overcome the tidal wave of public disgust at the Commission findings and The Coal Mines Act came into effect only three months after publication in August 1842. Women and girls were from then on banned from work below ground in mines, and boys had to be aged ten or older. Women and girls removed from the mines often bitterly resented their livelihoods being taken away from them. Many coal owners and colliery officials also felt the process should have been more gradually applied. The luckier ones – for instance, those losing their jobs in the Worsley pits – were even compensated to the tune of 2s a fortnight until they found alternative employment within the company's estates.

Women mineworkers had traditionally worn petticoats over trousers. Increasingly, many women and girls, now intent on sneaking below ground and often with the management turning a blind eye, were to be seen on the surface in more masculine caps, jackets and trousers. Seventy years after the 1842 legislation this style was still in place, particularly in Wigan, giving us the memorable image of the 'pit brow lasses'. The rest of Europe was slow in adopting female exclusion from the mines after 1842. Eventually, in 1886–87 and again in 1911, Parliament attempted to totally halt the employment of women and girls on the surface at collieries but was thwarted by successful protests and deputations to London by the girls, resplendent in their working outfits.

The general public had never really understood the lot of the miner and his wife. The large number of miners' wives who worked on the pit brow endured a seemingly ceaseless round of toil, being busy from dawn till dusk, struggling to win the never-ending war against dust and dirt. Husbands, sons or lodgers would tramp dust through the house, hot water for baths had to be prepared several times a day for those returning from the pit and meal times were staggered to fit the hours of the working men. On top of all this the children would have to be cared for. Many women (my grandmother being an example) took in washing or mending to supplement their meagre incomes.

Times were hard for mining families, but during the fairly regular disputes and strikes the situation was much worse, with women having to perform wonders to keep the family alive. Here the sense of community and unity amongst the women of the coalfields, historically very strong, came to the fore and ensured their survival. This was eventually seen in the great strike of 1984 where women's support groups emerged, the Lancashire Women Against Pit Closures movement being our local example. The memorable battle they fought against insurmountable odds was to be their last as the industry and its communities were pre-destined to be wiped out for purely political reasons. That the great British coal industry should be brought to an untimely end in this way has to be one of the greatest crimes ever perpetrated against the English working classes.

The few studies on the lives of women both in the mines and on the pit brow, published mainly over the last thirty years, have ranged enormously from sympathetic and accurate local history accounts by members of former mining communities to some misguidedly extreme 'academic' feminist and other studies of Victorian male perversions among the purchasers of pit brow women photographs and cards.

This study is aimed at anyone who, like myself, has either worked in mining, studied mining itself or its history and has the greatest respect for the men and women who have worked at our collieries. We can be proud that the Wigan area had the highest concentration of pit brow women in the country, women so full of character and typical Lancashire 'grit' that they became famous as a result, and we can be proud that such a high-quality photographic record has survived.

Alan Davies
Archivist, Wigan Leisure and Culture Trust
2006

one

The Early Days

The employment of women below and above ground in the Wigan area is sadly not particularly well documented within the borough archives. The fact that women and young girls were at work in local mines was not felt to be particularly unusual, shocking or controversial until well into the nineteenth century.

We can get a glimpse of evidence of female employment in the areas pits through church burial records. Very few churches indicated in burial records what the former occupation of the deceased was. Wigan All Saints Church of England did actually take the trouble to do this and first mentions a female mining death in the burial on 4 October 1641 of Elizabeth Higenson (probably Higginson) 'of the Woodhouses killed in a cole pit'. Richard Barron of Standishgate also died in the same accident, so it is probable that Elizabeth was loading and dragging away baskets of coal for Richard when either an ignition of methane gas took place at the coalface or a roof fall occurred.

The rapid expansion of the coal industry from the mid-eighteenth century onwards brought with it an increased death toll. 'Cannel miner' (cannel is a special type of coal once highly prized)

A female surface worker using a windlass to wind children down the mineshaft, an engraving from the Children's Employment Commission of 1842. The young boy and girl straddle the iron bar used to suspend baskets of coal. Shaft accidents were common at this time.

In 1842 women were not allowed to work as colliers, a role reserved for the men, but worked with them loading and riddling (screening) the coal at the coal face as in this engraving. They died from roof falls and explosions along with the colliers.

1842 Commission. A young girl putting in her maximum effort as she gets her head behind the coal cart in a difficult section of roadway.

Fig. 8.

Two views from 1842, the top showing a girl working virtually naked pulling herself along the roadway by holding onto the floor, the lower image showing a girl using the power of her back to 'thrutch' the cart out of some bad ground. Note the candle poorly illuminating the back end of the cart while she works in virtual darkness.

Lucia Greenhough, aged fourteen, was buried at Wigan All Saints on 1 December 1781. She was the daughter of John Greenough of Woodhouses, also a cannel miner, and may have died working alongside him hauling his coal away.

An article in *Gentleman's Magazine* entitled 'The Lancashire Collier Girl' describes conditions in the Orrell district in 1795. Here young Betty Hodson was at work down Halliwell's Fire Engine Pit along with father and brother, having started below ground aged nine and seven respectively. In the article it states:

> These little folks soon put their strength to the basket, dragging the coals from the workmen to the pit and by these efforts did the duty as it is called of one drawer. It is with pride that we make it known to little children that Betty and her brother at an early age cleared their parents seven shillings a week. Here was a treasure and satisfaction they were taught to feel by example and by the encouragement given to them.

In the 1842 Commission, Betty Harris of Bolton, depicted here, gave evidence of how difficult her job could be. At times she had to hold on to a rope to pull her sledge-mounted basket up inclines. The presence of the young assistant would not have helped much.

Women and girls working at the larger collieries pushed their baskets of coal from the coal face to the pit bottom on 'rolleys', mounted on simple cast-iron bar rails. This made the task easier so coal owners added another basket!

Left: *This lady had worked down Edge Fold Colliery, Worsley, until the Coal Mines Act of 1842 banned women and young girls from working below ground. Here she poses in the 1860s with her old harness, breeches, cap and hand-held candle. Worsley women were lucky to be compensated for the loss of their jobs and given work on the large Bridgewater estates.*

Below: *A leather and chain human harness of about the 1830s. They were found in the Old Gib Pit workings at Gibfield Colliery, Atherton, in the early 1950s.*

Betty's father was to die at the pit as a stone fell down the shaft and hit him as he was hooking a basket onto the rope. The feature continues:

> At her father's death Betty was between eleven and twelve years old and she continued in the coal pit in preference to throwing herself on the parish, as she was then, by her own labour, capable of earning a shilling a day. At her full strength she got 2/6d and at sixteen took her mother to live with her.

Eventually, Betty became exhausted in her late teens after her mother died and was luckily offered a job at Winstanley Hall, later becoming the cook. A description of her soon after starting at the hall shows the respect the public had for collier girls:

> … well looking, tall and slender in person, with grey eyes and bold countenance… but it is the boldness of honesty… she still maintains that fearless character of the miner whom no dangers can possibly daunt.

Wigan Coroners' Records

Although the Coroners Court was of ancient origin the records for Wigan have only survived from 1804 onwards with many lengthy gaps in the run. The first female recorded in the surviving run is Alice Oliver, 'accidentally killed in a coal pit' on 10 June 1838.

Mary Lowe fell down the shaft at Mesnes Pit (near the present-day park) on 14 June 1839. She may be our first actual record of a pit brow woman suffering a fatal accident but without a doubt many women and girls had died over the previous centuries. An amazing record for a woman working and living on the surface at a Wigan area colliery is that of Ellen Smith, born in a cottage close to the shaft and engine house at Gibfield Colliery, Atherton, in 1846. Her father worked as the winding engineman. After her father died when she was thirteen she worked at the pit for John Fletcher & Others (later Fletcher Burrows & Co.) from 1859 until 1921, sixty-two years' unbroken service.

The Children's Employment Commission (Mines) 1842

Before 1842 legislation was not in place to improve or safeguard the working conditions of those employed in the Wigan area mines. Virtually the single driving force to eradicate the employment of women and girls in the mines was to come from a most unlikely direction, a Tory MP Lord Anthony Ashley Cooper, seventh Earl of Shaftesbury. The findings of the Children's Employment Commission of 1842 have been recalled in many a publication or student project over the years, yet very few writers really give Anthony Ashley Cooper the enormous credit he deserves.

The Earls of Shaftesbury had a long tradition of maintaining a strong sense of moral responsibility. Anthony Ashley Cooper's illustrious ancestor, the third Earl of Shaftesbury, lived from 1671 to 1713. He was one of the most important philosophers of his day, and exerted an enormous influence throughout the eighteenth and nineteenth centuries on British and European discussions of morality, aesthetics and religion. The eldest son of the sixth Earl of Shaftesbury, Anthony was born on 28 April, 1801. Aged seven he was sent to boarding school and five years later transferred to Harrow. At the age of ten, Anthony was given the courtesy title of Lord Ashley. Harrow was followed by Christ College, Oxford, and at the age of twenty-five he was elected as MP for Woodstock, then under the control of the Shaftesbury family.

Lord Ashley began to take an interest in social issues after reading reports in The Times about the accounts given to Michael Sadler's Committee investigating child labour. Lord Ashley wrote to Michael Sadler offering his help in his campaign for factory reform. When Michael Sadler was defeated in the 1832 General Election, Revd George Bull, the Evangelical curate of Bierly near Bradford, asked Lord Ashley to become the new leader of the factory reform movement in the House of Commons. Ashley's critics claimed that he took up the factory question 'as much from a dislike of the mill owners as from sympathy with the mill workers.' Lord Ashley agreed to George Bull's request and in March 1833 proposed a bill that would restrict children to a maximum ten-hour day. On 18 July 1833 Ashley's bill was defeated in the House of Commons by 238 votes to ninety-three. Although the Government opposed Ashley's bill, it accepted that children did need protecting and decided to put forward its own proposals.

1833 Factory Act

The Government's 1833 Factory Act was passed by Parliament on 29 August. Under the terms of the new act, it became illegal for children under nine to work in textile factories, whereas children aged between nine and thirteen could not be employed for more than eight hours a day. The main disappointment of the reformers was that children over thirteen were allowed to work for up to twelve hours a day. They also complained that with the employment of only four inspectors to monitor this legislation, factory owners would continue to employ very young children.

Children's Employment Commission

In 1840 Lord Ashley worked with others on establishing the Children's Employment Commission. Its first report (on mines and collieries) was published in 1842. Sections were headed as follows:

1. Ages at which children and young persons are employed in coal mines
2. Sex: employment of girls and women in coal mines
3. Number of children and young persons in coal mines
4. Hiring of children and young persons in coal mines
5. State of the place of work in coal mines
6. Nature of the employment in coal mines
7. Hours of work in coal mines
8. Night work in coal mines
9. Meal hours in coal mines
10. Holidays allowed to children and young persons in coal mines
11. Treatment of children and young persons in coal mines
12. Accidents to which children and young persons are exposed in coal mines
13. Wages of children and young persons in coal mines
14. Influence of employment in coal mines on the physical condition of children and young persons

The report caused a sensation when it was published. The majority of people in Britain were totally unaware that women and children as young as six years of age were being employed in mines. Lord Ashley piloted the Coal Mines Act 1842 through the House of Commons. As a result of this legislation women and girls and boys younger than ten were prohibited from working underground.

The evidence collected by the commissioners in the various coalfields gives an amazing insight into the cruel reality of life for the men, women and children in mining communities in the early 1840s. The verbatim evidence includes local dialect, some of which was unique to that particular mining community, and which must have totally puzzled the elected member for East Sussex!

Graphic Illustrations

The report was the first of its kind to include engraved illustrations, which for the already shocked Victorian reader added an unbelievable visual dimension. Images of muscularly deformed naked young girls dragging coal carts along low tunnels were to be instrumental in speeding up the Parliamentary route to a change in legislation. The engravings were reproduced in the national newspapers, standing out at a time when pictures in newspapers were rare, and images of bare-breasted females even rarer! The illustrations, most of which were to be found in the Lancashire sections of evidence, also give us the first accurate glimpse into working conditions and mining 'technology' below ground.

The engravings appear to have been accurately sketched in a cold manner, without any form of artistic embellishment or licence. Perhaps the artists felt a sense of moral duty to record accurately the horrors before them, which they themselves were probably seeing for the first time.

The Evidence

This was gathered by the sub-commissioners below ground and amongst the mining community on the surface, in local pubs, for example, a well-known haunt of most colliers. Some smaller coal owners who knew full well that their workforce was enslaved in terrible conditions were very uncooperative, their mineworkers giving little useful insight into conditions. Larger concerns such as the Duke of Bridgewater's massive mining complex based around the underground canal system accessed from Worsley were the opposite. Even though they employed over 140 women and girls, they readily supplied statistics and allowed workers to openly speak to the sub-commissioners.

There was a great deal of suspicion of the motives behind the gathering of evidence. Mineworkers felt that revealing their terrible working conditions and immoral practices to an audience outside of their close-knit communities would lead to job losses, colliery closures or prosecution. As mining wages were often higher than mill wages the workforce often felt it worthwhile to keep quiet, accept their lot and hope for the best.

The evidence that follows only scrapes the surface of the scope of the full report. As the wider Lancashire coalfield was under investigation, the present Wigan Borough area is very limited in coverage, more so if we narrow our search to women and girls' evidence. The testimonies of women and girls from nearby coalfields give us an insight into what conditions in the borough must have been like.

Ellen Taylor was interviewed at Bundell's Pemberton Colliery in 1841:

What age are you? Eleven years old.
How long have you worked in the pits? About four years.
Can you read or write? I can read, but I cannot write.
Do you ever get beaten? No, not I'th pit, but my mother beats me sometimes when I'm naught (nowt or naughty).

Children were given tasks an adult of today would never accept. An astonishing piece of evidence comes from Benjamin Berry, aged fourteen, working at Patricroft Colliery, near Worsley and includes evidence of the girls' arduous tasks:

What is the greatest length you have known a drawer bring a wagon (during a shift)? I have known two drawers, a lad and a lass, draw 800 yards on the level, with rails, ten times (16,000 yards) and 360 yards up and down without rails twenty times (14,400 yards); total 30,400 yards or 17½ miles nearly.

Many ramblers today would be proud to walk 17 miles in the Lakes, but just imagine dragging loaded and empty tubs in the mine for ten hours for the same distance!

On top of enduring terrible working conditions, the women and girls in the Wigan area featuring in the report often returned home to shocking domestic lives. William Harrison was the Relieving Officer for Orrell, Pemberton and Upholland and several other townships. In his evidence of 12 May 1841 he replies to the following question:

Have you ever noticed the domestic condition of the children of colliers? As regards their habits, some of them are very respectable, but the great body are quite the reverse, and very disorderly in conduct, and they make bad use of their money; they are usually filthy in their habits, they will wash their neck, face, and hands, when they return from work, but they scarcely ever wash their bodies, very rarely indeed.

Do they (colliers) attend places of worship? Very few of them do, some of course do, but very many do not, they would rather be in the alehouse than the church, especially of the Saturday night and Sunday, they are there as much as they possibly can be without being detected; fights are frequent, and breaches of the peace, but since we had the police we are now more quiet than formerly.

Have you ever observed the moral condition of the colliers? It is not uncommon for four or five of them to sleep in one bed, and sometimes two or three beds in one room; the girls and boys sleep in the same room, and I have known cases where they have slept in the same bed. I have known them sleep together until the boys and girls were fourteen or fifteen. They are very much demoralised in their habits; bastard children are thought nothing of, it scarcely makes any difference on a young woman's character in this district, if they get a child they think they should go to the parish with it.

Mr Birchall, the Relieving Officer of the Hindley District, including Ashton, Aspull, Blackrod, Haigh, Hindley, Ince, was asked:

Have you remarked drunkenness as usual amongst them? Yes, I have; I consider they rob themselves of necessities to get to the pothouse and indulge in that vice; they leave their families in a state of destitution, and in the absence of the parents the children get into all kinds of mischief and bad company. I have frequently seen quite young boys at the public house with their fathers, smoking, singing, and drinking like the rest.

Mr Latham was the hard-pressed Chief Constable of Wigan. He had been a resident for forty-two years and amazingly was still asked the following question:

In that time have you had the opportunity of noting the character and habits of the colliers in this neighbourhood? Yes, I have; and all classes of working men in this district, excepting the lower classes of Irish, who are probably as bad, the colliers are the most turbulent and riotous; their language is of the most blasphemous description, and the children follow their example; the name of Christ is always in their mouths; their constant oath is, 'by the heart of Christ I'll tear thy liver out.'

Is it considered discreditable for a man to be seen drunk? No, it is considered no disgrace to be seen drunk and disorderly; every pay night they come to the town, and they are drunk and disorderly upon these occasions; their wives usually accompany them, and leave the children to fend for themselves.

In visiting the houses of the colliers, are scenes of riot and debauchery frequent? In visiting houses lately, a policeman reported that he saw the wife on the floor in a state of beastly intoxication, and the children half naked, sitting on their heels round the fire.

The main task for women working below ground was to shovel up coal the collier had brought down with his pick, riddle it through a sieve or mesh and load it into either a basket or a wooden tub on wheels. Using a belt and chain they then had to drag or push the container to the pit bottom or a loading point. The next piece of evidence, although from a woman working down a Bolton pit and not a million miles from Wigan Borough, is probably valid for the majority of women working in British mines.

Betty Harris, aged thirty-seven, was a drawer (pushing and pulling coal tubs and baskets) in a coal mine at Little Bolton:

I have a belt round my waist, and a chain passing between my legs, and I go on my hands and feet. The road is very steep, and we have to hold by a rope; and where there is no rope, by anything we can catch hold of... I am not as strong as I was, and cannot stand the work as well as I used to. I have drawn till I have had the skin off me; belt and chain is worse when we are in the family way.

Patience Kershaw, aged seventeen and also from Bolton describes a working routine where the job itself was not the only cause for concern:

I hurry the corves [pull or drag wicker baskets full of coal] a mile and more underground and back; they weigh 3 cwt [20 cwt to a ton]... the getters [the men at the coal face] that I work for are naked except for their caps, sometimes they beat me if I am not quick enough [in supplying them with empty baskets].

Betty Wardle of Outwood, near Radcliffe, recalls her time in the pits. A modern-day mother will find her evidence virtually unbelievable:

Have you ever worked in a coal pit? Ay, I have worked in a pit since I was six years old.
Have you any children? Yes, I have had four children; two of them were born while I worked in the pits.
Did you work in the pits when you were in the family way? Ay, to be sure. I had a child born in the pits and I brought it up the pit shaft in my skirt.

Are you quite sure you are telling me the truth? Ay, that I am; it was born the day after I were married, that makes me to know.

Entrepreneur Mary Glover, working at a pit near Ringley was asked:

How are you dressed when at work in the pits? I wear a shift and a pair of trousers when at work, and I always will have a good pair of trousers. I have had many a 2*d* given me by the boatmen on the canal side to show my breeches.

Rosa Lucas, nearly eighteen years old, at Mr Morris's, Lamberhead Green, Wigan, 19 May 1841:

You are a drawer, I believe, when at work? Yes, I am.
Where do you work? At Mr Morris's. I used to work at Blundell's.
What age were you when you first began to work in the pits? I was about eleven I think.
Do you work at night in Morris's pits? Yes, when I was able to work. I worked one week in the daytime and the next at night, the same as the drawers did.
Are there any children in the pit where you work? Oh, yes, both little employed in the pits and big, some not older or bigger than him [pointing to a little boy of six or seven years old]; they put them to tenting air doors.
What hours do you work? I go down between three and four in the morning and sometimes I have done by five o'clock in the afternoon, and sometimes sooner.
Have you any fixed hour for dinner? Yes, we have an hour for dinner during the day in the daytime, but we don't stop at night.
When you are working the night-turn, what hours do you work? I go at night at two o'clock in the afternoon, and sometimes three. I come up it will be about three o'clock in the morning, and sometimes before.
You have no regular times for meals at night? No, we never stop at during the night.
Do you find the work very hard? Yes, it is very hard work for a woman. From over work. I have been so tired many a time that I could scarcely wash myself. I was obliged to leave Mr Blundell's pit, it was so hot, and my work was a deal harder; I could scarcely ever wash myself at night, I was so tired; and I felt very dull and stiff when I set off in the morning.
What distance did you draw? 23 score yards in length.
That is 460 yards each way, or 920 yards? Yes.
How many times had you to draw this distance? 16 and sometimes 18 in one day. [Taking 16 times, she would have to draw 14,720 yards daily.]
Have you ever had many accidents besides the one you are now suffering from? Yes, I had once a great big hole in my other leg. I thought it was the water that did it, for I was working in a wet place then.
Are Mr Morris's pits dry? Yes, very dry.
How did the accident happen you are now suffering from? I was sitting on the edge of a tub at the bottom, and a great stone fell from the roof on my foot and ankle, and crushed it to pieces, and it was obliged to be taken off.
Have you ever seen the drawers beaten? Yes, some gets beaten. Tuity gets beaten nearly every day.
What do they beat her with? A pick-arm.
What do they beat her for? I suppose it is for 'sauce'; she has a very saucy tongue.
What age is she? She is twenty-three years old.
What is your father? He was a collier, but he was killed in a coal pit. I go past the place where he was killed many a time when I am working, and sometimes I think I see something.

Margaret Winstanley, drawer, aged twenty-four, at Mr Thickness's Colliery, Wigan:

In whose employ are you? Mr Thickness's.
How long have you been a drawer? For fourteen or fifteen years.
Are you married? Yes, I am, and this is my first child, it is eleven months old.
Have you been at work since the birth of your child? Oh, yes, I went back in a month after it was born.
How does it happen that you are not at work today? Why, the pit I work in is very wet – the water is half a yard deep in some places; and my husband has gone to his master to see if he won't put him in a dryer place, for he cannot work where he is.
Have you good health? Not lately. I have been very ill. I lay in my bed several days, and my child was ill, too, from my working in the wet so much.
Why do you not change your place of work? My husband is bound for eleven months to Mr Thickness; and his master says he must work either there or go to Kirkdale [i.e. the gaol], which

One of our earliest photographs of Wigan area pit brow women, seen at John Morris's Rose Bridge Colliery, near Ince, in 1865. One of a pair of images produced for a hand-held stereo viewer.

he pleases. My husband has worked in wet places for many a year; sometimes he has worked up to his knees in the wet, and does now when he is at work. When I am drawing for him my clothes are all wet through.

Will not his master allow him to give up this agreement? No, he has lent him money, and he will have to work for eleven months before it is paid off.

Have you ever been injured in the coal pit? No, not to signify. I have been hurt, but colliers don't make any account of being hurt, unless their bones are broken. My brother had his leg hurt by the roof falling, and his leg has been taken off; he is getting well now very nicely.

The Children's Employment Commission was published in May 1842. The large number of coal owners in the House of Lords opposed the removal of women and young girls from the mines as expected. General public outrage at the revelations meant that many of the workplace practices or typically low moral habits could no longer be condoned. The Coal Mines Act came into effect only three months after publication in August 1842. Women and girls removed from the mines often bitterly resented their livelihoods being taken away from them in one fell swoop.

An early and rather sad image of a pit brow woman at Shevington Colliery in 1867.

Many coal owners and colliery officials also felt the process should have been more gradually applied. Those lucky women and girls losing their jobs in the Worsley pits were compensated to the tune of 2s a fortnight until they found alternative jobs. Other women and girls were lucky to be given agricultural or service work on the estates of the coal owners or drifted away to other industries, such as the burgeoning textiles industry, or even cockling in Morecambe Bay.

Mr Peace, agent for Lord Balcarres at Haigh, had stated to the Commission that his Lordship was anxious to discontinue the employment of females in his collieries in the neighbourhood of Haigh, Aspull and Blackrod, but the system had been carried on for so many years and there were so many females employed in them that it would be impossible to dispense with them all of a sudden.

He stated:

> Unless the measure is very slow and gradual in it's operation the immediate misery it will produce will be greater than the vice and immorality it is certainly well calculated to check.

Cambridge diarist, barrister and poet A.J. Munby stands alongside one of the many pit brow women he had the greatest respect for, Ellen Grounds, aged twenty-two. This was in a Wigan studio on 11 September 1873.

Some women around Wigan resorted to dressing as men and continued working for years after the act came into force. Hannah Hathaway was discovered after being killed in a roof fall at a Standish pit in 1845. An enquiry found other women had been working at the pit.

In 1846, once more in relation to the Haigh collieries, Peace wrote:

> We are suffering great loss and inconvenience through the employment of women by the proprietors of collieries on all sides of us. About a dozen men have left our Aberdeen Colliery and moved to Burgh Colliery near Chorley because there, their wives and daughters are allowed to work down the pit. I am reliably informed that upwards of 30 women are employed in one pit at that colliery. It is the property of Mr Hargreaves of Bolton. Our Aberdeen Pit is in consequence of this being carried on with about two thirds of its full complement which causes us great loss; the expense of keeping open, ventilated and drained the workings as well as engineers and other numerous attendants paid by day wages being nearly the same for a small produce of coal as for a large. At Mr Wood's colliery near our central pits in Haigh, he is knowingly employing great numbers of women, whom I frequently meet returning from his colliery to their residence in Haigh and who frequently ask me to allow them to resume their labours in our own pits, seeing that it would be much more convenient and agreeable to themselves to work in our mines which are better ventilated and generally in better and safer condition than many of those in the neighbourhood, and seeing that Lord Ashley's Act is openly disregarded by our neighbours on every side.

Despite colliery owners flaunting the law all around him taking on his former female employees, Lord Crawford adhered strictly to the changes in legislation and did not allow women below ground. By 1860 work for thirty ex-Haigh pit women was being provided at Sumner's Brewery, close to Haigh St David's church.

Nineteen years after the Coal Mines Act of 1842 the general public were still relatively ignorant of the problems facing the industry. C. Collier in his *Gatherings From The Pit Heap of 1861* stated that:

> ... of miners the mass of mankind knows no more than if they were Hottentots;- born, bred and buried, for the most part out of sight of the highly civilised and educated people around them.

To people not acquainted with mining work, the former working lives of the women and girls had appeared unbearably harsh and needed to be controlled. In reality, for many of these fit, healthy and strong women the earnings were vital and the work was all they had known since childhood. They were accustomed to its rigours. Many openly stated that they enjoyed it!

two

The Pit Brow Lass Emerges

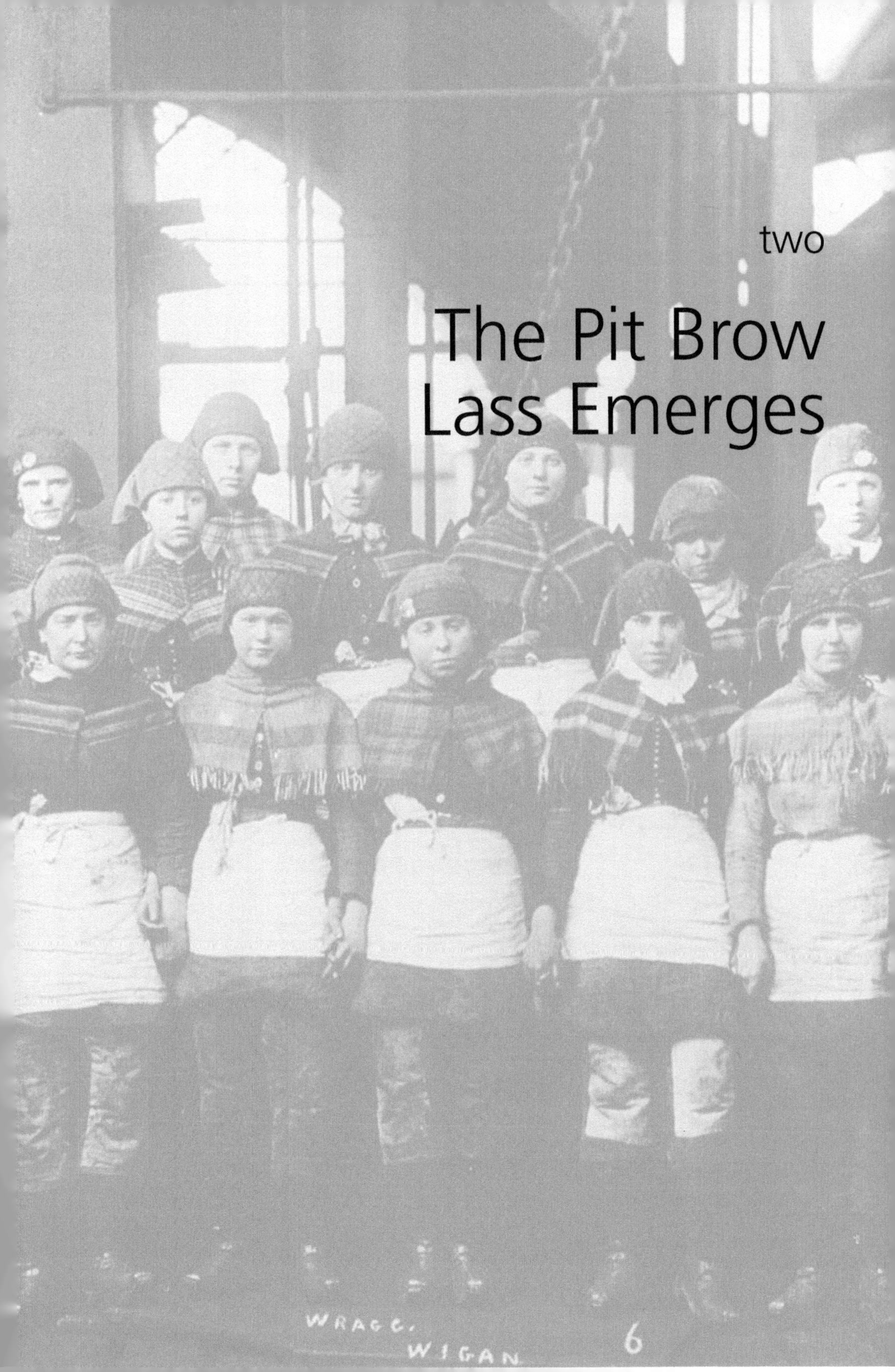

Wigan area female mineworkers had traditionally worn petticoats over trousers. Those intent on sneaking below ground, aided by 'blind' colliery management, were increasingly to be seen in more masculine caps jackets and trousers. Seventy years after the 1842 legislation this style was still in place in the Wigan area, yielding the famous impression of the pit brow lasses.

Not all Wigan area collieries immediately welcomed women onto the surface. Men who had been injured below ground had traditionally secured the less arduous jobs on the surface. Some colliery owners felt the pit was not the place for a woman to be seen, surrounded by ill-behaved and foul-mouthed men. Others were more than happy to take on women and girls who they regarded as good, strong, reliable workers, and they preferred women from mining families who were used to the manners of miners and pit surface workmen.

We are very lucky that Cambridge academic A.J. Munby, civil servant, barrister, diarist and minor poet, took a special interest in the pit brow women of the Wigan area. There are those who have totally and misguidedly interpreted his interest as a perverted one. He kept a detailed diary of his encounters with the women and gives a fascinating insight into their manners and dress. He also had photographs produced of the women by local photographers. Through Munby's diaries we have a unique link with the women of the 1842 Commission as some of the older women he met would be former underground workers. Extracts follow:

An early and atmospheric studio portrait of a miner and pit brow woman by Louisa Millard of Wigan, dated 1869.

Selected extracts from the diaries of A.J. Munby (1828–1910)

Friday 19 August 1859

Got out of the train at Hindley, where the Wigan coal district begins. The first pits I came to were of the Kirkless Hall group, three close together, and about a quarter of a mile from the station. Those black, nondescript creatures pushing the wagons along the embankments would not be noticed by travellers on the line, they would pass for men, but I recognised them at once as my stout hearted friends, the Lancashire collier girls.

The costume of these girls and women is always the same, and a good useful one it is. A hooded bonnet of padded cotton, pink blue or black. A blue striped shirt, open at the breast, a waistcoat of cloth, generally double breasted – but ragged and patched throughout. Fustian or corduroy or sometimes blackcloth trousers, patched with all possible materials except the original one, and stout clog shoon, brassclipped on their bare feet. Round the waist is tucked a petticoat of striped cotton, blue and black, rolled up as a joiner rolls his apron: it is never let down, and is perfectly useless – only retained as a symbol of sex.

At the first of the three pits I found four women, all young, one was standing with her hands in her pockets, and another sitting dangling her legs on the edge of a railway coal truck, waiting for coal's from the pit's mouth to fill it. The third girl I found behind a great bank of coke, digging it down with a spade ready for the corves.

While talking to her, a train of empty corves started down the slope towards us from the pit's mouth: the fourth girl who was in charge of it took a flying leap as she set the train a going, and stuck on to one of the corves till they reached the bottom where we were: then jumped off, a straightway seized her spade and fell to digging without a word. She was a pretty brunette of eighteen, strong and healthy: her clothes, even her coarse flannel trousers, were in good condition: and dirty as she was, she had been woman enough to stick a bunch of half ripe wild cherries in the side of her grimy bonnet.

Saturday September 1860

I went to a neat row of stone cottages just beyond Frank Holme's (a public house at Springs Bridge, near Wigan) and knocked at the door, which stood open at the centre. I went in, and found a room comfortable in a rough way, but untidy, it being cleaning night. Before the blazing fire sat a boy, grimy and in pit clothes, resting after his work: and a bony sallow girl from a factory was eating her supper at a dresser.

These were Jane's brother and sister. Her mother came in directly: a stout loud voiced woman, hot tempered evidently, but good humoured and kindly in a blunt fashion; comely also and neat enough.

She asked me to sit down; I explained about the picture [he had come to buy a photograph of Jane in her trousers] and she was going to produce it, when shouts and trampling of clogs soon were heard outside, the door was driven open, and in burst the two wenches, Ellen Meggison and Jane, shouting and tumbling over one another like lads at a fair. They were both of course in their pit clothes, and as black as ever; and their grimy faces were bathed in sweat, for they had been running home all the way.

The mother instantly began to rate Jane soundly for staying at the alehouse and being out so late (it was now 7.30): 'Is that the way for a respectable young woman?' thundered she

Studio image of a pit brow woman by Louisa Millard of Wigan dated 1869. In her left hand she is holding a group of pit tub tallies. These were attached to tubs by colliers below ground to identify them when they arrived on the surface to be weighed. The men were paid by the amount of coal they produced.

Studio portrait of pit brow women by Wragg of Wigan, c.1890, the women apparently still retaining their pit dirt on their hands and faces. A number of portraits of pit women were produced by Wragg with the same basic props arranged slightly differently.

– evidently not merely because I was present. The girls shouted in reply that they had to do overwork and to wait for their wages: and the hubbub subsided, and Ellen flung herself into a chair, and Jane leaned against the drawers, panting and wiping the beaded sweat – and with some of the blackness – from her red face, with the end of her neckchief.

It was a singular group – the quiet coaly boy by the fire, the sallow pale sister and stout clean mother, and then those two young women in men's clothes, as black and grimy as fiends and rough and uncouth in manners as a bargee, and yet, not unwomanly nor degraded. They soon became quiet, after the first burst: and their talk had nothing flippant or immodest in it. Jane sat down to her supper of Irish stew; scooping the potatoes out of the bowl with a wooden spoon, and holding the meat in her black fingers while she tore it from the bone with her teeth.

Her mother and I meanwhile stood and looked at her – she eating away unconcerned and hungry – and remarked what a fine healthy wench she was, and how she was not seventeen till next month, and so on. And yet this collier girl of seventeen is ten times more robust and womanly than her elder sister the factory girl... At last I had the picture (which was not a good one) produced, gave Jane the shilling for it, and sixpence to Ellen for 'a gill of ale' and said goodnight...

Here I am sorry to admit that as I left the cottage, Ellen started up and ran after me, dragging Jane with her, to beg that I would come to Frank Holme's and treat them to some gin. When I flatly refused, however, they both retired quietly, saying goodnight; Ellen going home to her grandmother's cottage further on.

An early studio shot of around 1870, the pit women wearing distinctive headscarves. Probably from the studio of Cooper, Wigan, who was based at various sites in town between 1853 and 1892.

No. 6 in the Wragg of Wigan series of around 1886, the colliery unknown.

A superbly detailed study of a young pit brow girl from the Wragg series. As it is No. 18, this could mean that as many as twenty such images may have been produced.

The Wragg series; this lady is thought to have worked for Lamb and Moores Newtown Colliery. Note the typical women's clasp-fastened clogs.

The Wragg series, No. 3. Note the massive 'number one' flat-bottomed coal shovel; flinging a full load from one of these would take some effort, even for a man. Behind the shovel stands the 'riddle' or sieve for grading coal, anything passing through the riddle being termed 'slack'. The wicker basket and water can contrast well with today's working woman's designer handbag, mobile phone and ready packed sandwiches!

A smiling Wragg studio group of brow girls around 1886, patterned scarves in evidence this time.

Wednesday 19 August 1863

I went out soon after nine, to Dugdale the photographer. Found he had lately taken a good picture of one Jane Horton, aged 19, a collier at Patricroft (Ince Hall) and formerly a factory girl. She was a niece of Dugdale's next door neighbours, a very respectable woman, who offered to send for her at once, if I liked. Apropos: I visited this and one or two other 'cartes de visite' of girls in pit clothes hung up in the street by the photographers; one of whom in Clarence Yard told me they often come to her to be taken thus clad, and rather like to be exhibited.

In the following section Munby passes through Windy Arbour, then downhill to Samuel Stocks Leyland Green Colliery and Simms Lane End.

Friday 17 March 1865

Fine sunny morning: afterwards cold and cloudy, with keen east wind. I went out at nine to the photographers; to Dugdale, who has at length taken a group of pit wenches at a colliery near Tayleurs, towards Gathurst to Cooper, who sells beer as well as photos, and who said he had sold hundreds of cartes de visite of the collier girls, who has already told me the same story: and to Little, an inferior party near the station. A case of his photos hung on view in the main street, and amongst them were several portraits of pit girls in costume. As I looked at them, two young women in female clothes and with shawls over their heads came up and looked also.

The long-established Cooper studio of Wigan produced images of pit brow women but without the scenic props used by Wraggs. Here, around the 1870s, a lady poses with just a cob of coal nearby to remind you of her job.

Cooper studio again around 1875. Four hardworking women manage a smile, their muscular arms hinting at the arduous nature of their work.

A portrait from about 1880 of an older pit brow woman, Margaret Hunter, who worked at Ince Hall Collieries. Looking to be in her late sixties, she may have actually worked below ground until 1842 when women ceased to be allowed down the pit. Fear of the workhouse might mean she worked until her late seventies.

Margaret Fairhurst of Ince Hall Collieries around 1880. She is wearing men's lace-up type clogs.

'Why' said one 'yon's Walsh Mary Ann!' and so it was; Ann Morgan of Hindley. 'And that,' she went on, pointing to a picture of a fine comely lass in loose shirt 'is Jane Underwood, that worked at Pigeon Pit; and that' – a good looking robust woman leaning on a spade – 'is King's greaser; Mary they call her; she greases the railway truck wheels. 'So you have worked on pit brow'? I asked. 'Aye, many a year!' said the girl. Just then an Officious Party, one of a small crowd which had gathered around, thought it well to explain what those trousered figures were. 'Them's women', he said, 'they're not men'. Men indeed!

All the four photographers said they sold these photos chiefly to commercial travellers, who buy them as 'curiosities'. Many strangers passes their remarks upon 'em', said Mrs Little; 'and some considers as it's a shame for women to wear breeches, and some takes it for a joke like.' Just so: some are sentimentalists, some sensualists: none rational and serious.

I left Wigan on foot at 11.45 to walk to St Helens and see the pits en route.

I was now about 6 miles from St Helens and it was 2.30pm. A respectable old man at the coal office on the brow advised me not to go via Billinge, which lay out of sight to the right: there are pits that way, said he, but more if you go the longer route by Blackleyhurst: such as those yonder, of Sammy Stock's. So I took his advice…

Going thence down the line, I came to a siding, running up the next pit, No 3. Here in the cutting I found a girl lounging about with her hands in her breeches-pockets, whistling. She turned out to be the points-woman, whose duty it is to mind the rails whenever a train of coal wagons goes by. I was struck by her brisk and easy air, and stopped to talk with her, and she seemed glad enough to have someone to talk to. Her name was Margaret Roughley, and her age 17; she has an elder sister working on the pit brow close by. This Margaret was a well grown girl: her collier bonnet was tilted over her eyes, which sparkled under lashes thickly clogged with coal dust: her face was very black, but also singularly expressive and intelligent: her arms were bare: she has a woollen comforter round her neck; a loose patched shirt, looking very thin and cold; a short baglike apron of sackcloth; short fustian trousers, only reaching to the calf; grey stockings, and big clog shoon, whose iron soles were turned up at the toe like a Chinaman's boot. Had she no coat? She had not. I asked: it must be very cold, waiting about in this cutting all day. No, but the Gaffer's very kind – he lends me his coat when it rains: and besides, she said, Aw don't stop here all day: Aw 'us wagons to fill wi' slack; 20 wagons a day sometimes; you'd not think it was idling if yo'd got that to do! And Margaret laughed a boyish laugh, and showed her white teeth, like Irish diamonds set in black bog oak. Then, as there were no trains coming, she leaned her back against the earth of the cutting, crossed her legs easily, tucked her hands under her 'barmskin' and looked straight at me, talking away for ten minutes as gaily and freely as if she had always known me. She spoke the broadest Lancashire, and spoke it so fast that I scarce follow her: e.g. 'Aw'd goo sairce'moo'r lemma' (service if mother'd let me). She didn't dislike this work, however: and although now she only earned 1/- a day while the broo wenches earn 1s/2d she hoped soon to be 'raised' from 10p a day already. She could not read: yet she was one of the brightest and most interesting girls I ever saw: as sharp and lively as a London street boy, yet with nothing impudent or unfeminine about her. True, her dress and manners were those of a lad; she jumped and ran, and hitched up her trousers, like one who knew nothing about petticoats, but in all her words and ways there was an artless simplicity and truthful frankness that was thoroughly girlish. She had humour too, and wit beyond her years; she was always saying something comic and then merrily laughing at her own conceit: she laughed at her grotesque attire: 'such queer old clothes as we wear' said she 'ragged ones that folk gie us – our brothers old coats and britches – anything does to work at pit in: but we wear our breeches always, yo know, 'cept Sundays – and nice and warm they are too!' and she threw her left leg over her right as she spoke. 'Eh!' she went on 'bud yo will laugh at thoose wenches at Two Pit – they're all such rags – ho ho!' and she laughed again.

Strange creature... A black ignorant collier girl of seventeen, in breeches and clogshoon, and yet winning and loveable, and clever enough to repay a deal of teaching. When I said I was going to see the other wenches 'Aw'll coom wi' ye' she cried, starting up 'and see hah they're getting on' – and we went. On the way, she remarked that she would look queer if she were 'drawed aht' just as she was. Why, did you never see them done? I asked. No, she said – Why, did you never see them done? I asked. No, she said – Why, they wouldn't do it would they? To be sure they would, said I, and straightway showed her one of my Wigan photos. The girl stared at it a moment and then shouted aloud with wonder and delight, and insisted on taking the picture into her black fingers, pleading that she wouldn't hurt it. 'Polly' she hallooed to her sister on the brow – 'coom quick! Here's a wench drawed aht in her pit claes – eh, Lord just look!' and all the other lasses, 5 in number, came running down to see.

Two had been thrutching corves, and 3 filling trucks: they were all stout robust young women of twenty or so, in trousers and thick jackets or coats. One had worked in a factory, but liked pit better. They looked with grave wonder at the photograph, till some man from the brow called – 'Polly, wench, bring thy spade!' and then they ran back to work. And my Margaret, who had more life and soul in her than all the rest had, cried out to me 'Goodbye! Aw'm going to warm myself!' and stamping and folding her arms, darted off at a smart run and disappeared.

... I went down, and on the No 1 pit close by. Here were about 9 young women working as at No 2 and equally stout and healthy. As I left them, a woman in petticoat and fustian trousers came out of the engine house in a leisurely manner, her hands in her pockets. She had a very striking face: aquiline features, a strong jaw and bold chin, and hazel eyes so large and keen that to meet their gaze was like breasting the blow of a mountain breeze. She said she was 37 years old, but looked nearly ten years younger: she was strong and square built, but not coarse nor large. And who is this beauty, whose face begrimed as it was, was so handsome and powerful and expressive. Why, she was a mere quadruped. She had been brought up in service, she said: but of her own accord had left that calling, and gone down a coal pit, about 15 years ago, to work as a drawer. Of course she went as a man; dressed in men's clothes and passed as a man; but she liked it, and liked the work. Did she draw with the belt and chain? 'Yes', she said: 'I was harnessed to the corves, with a belt round my body and the chain between my legs, hooked on to the corves'. And did not the harness hurt you? 'No, my breeches kept the chain from hurting my legs'. And you went on your hands and feet, just as a horse goes on four legs? 'Yes, just the same', she said simply: 'and the roads was rough – there was no rails when I draw'd; it was over my wrists in mud, often. I used to draw the corves 200 or 300 yards' she went on, 'I could do it easily: I don't mind going on my hands and feet for that distance, and more, oh no! And it did not make my hands very hard, using them as if they was feet; nor harder than they are now'. All this the handsome brighteyed woman told me with quite unaffected candour. 'I liked it', she said with emphasis: 'but when I'd been working down a month, they found out I was a woman, and I was turned out: and since then I've worked on pit brow and worn breeches, as I'm doing now'.

I was going to ask the name of this *mulier formosa* (beautiful woman) when I was interrupted. On the brow I had seen afar off a big rough looking man in a hat: unique phenomenon. And now this same man came suddenly up unawares, placed himself before me, and said with loud voice and angry tone to me 'What may your business here be? Go off!' I stared, and declined to tell my business without knowing who my questioner was. 'What's that to you? He cried: 'you're not welcome here: go off!'.

But as I did not go, he, after more reproaches, added, hoarsely 'You want to know my authority: well, I'm the proprietor of this colliery: and I don't want you here: and I tell you to

FEMALES AS THEY WORK AT THE PIT BANKS.
PHOTOGRAPHED FOR SALE AT WIGAN, IN LANCASHIRE, 1863.

At the Leeds conference of The National Association of Coal Lime and Ironstone Miners in 1863, the role of female workers on the pit brow was discussed. This engraved frontispiece worked up from photographs possibly by Cooper was included in the published transactions, entitled Females As They Work At The Pit Banks, Photographed For Sale at Wigan in Lancashire 1863.

go off!' and he clenched his fist. I then apologised for having intruded, adding that if I had been properly addressed I should have done so at first.

To which Sammy Stock (for it was he) replied, 'I don't want your apologies: go off!' and turning to the women, who had stood by silent, he added roughly 'Go to your work!'

She went, meekly enough: and I also took my way. It was worth remarking that this, the only act of rudeness that was ever offered to me at a coal pit, came from the master, and not from the workpeople, male or female.

Above: The Pictorial World, *18 April 1874, showing female 'trimmers' levelling the coal out on a main-line wagon. Mesnes Colliery, Wigan.*

Opposite: *Engraving from* The Pictorial World *of 18 April 1874 showing pit women at Charles Turner's Mesnes Colliery, Wigan, moving coal down the coal shoots towards the cart waiting to be loaded.*

The Pictorial World, *18 April 1874, portraying the classic Wigan pit brow girl with riddle and shovel. From a photograph taken at Mesnes Colliery.*

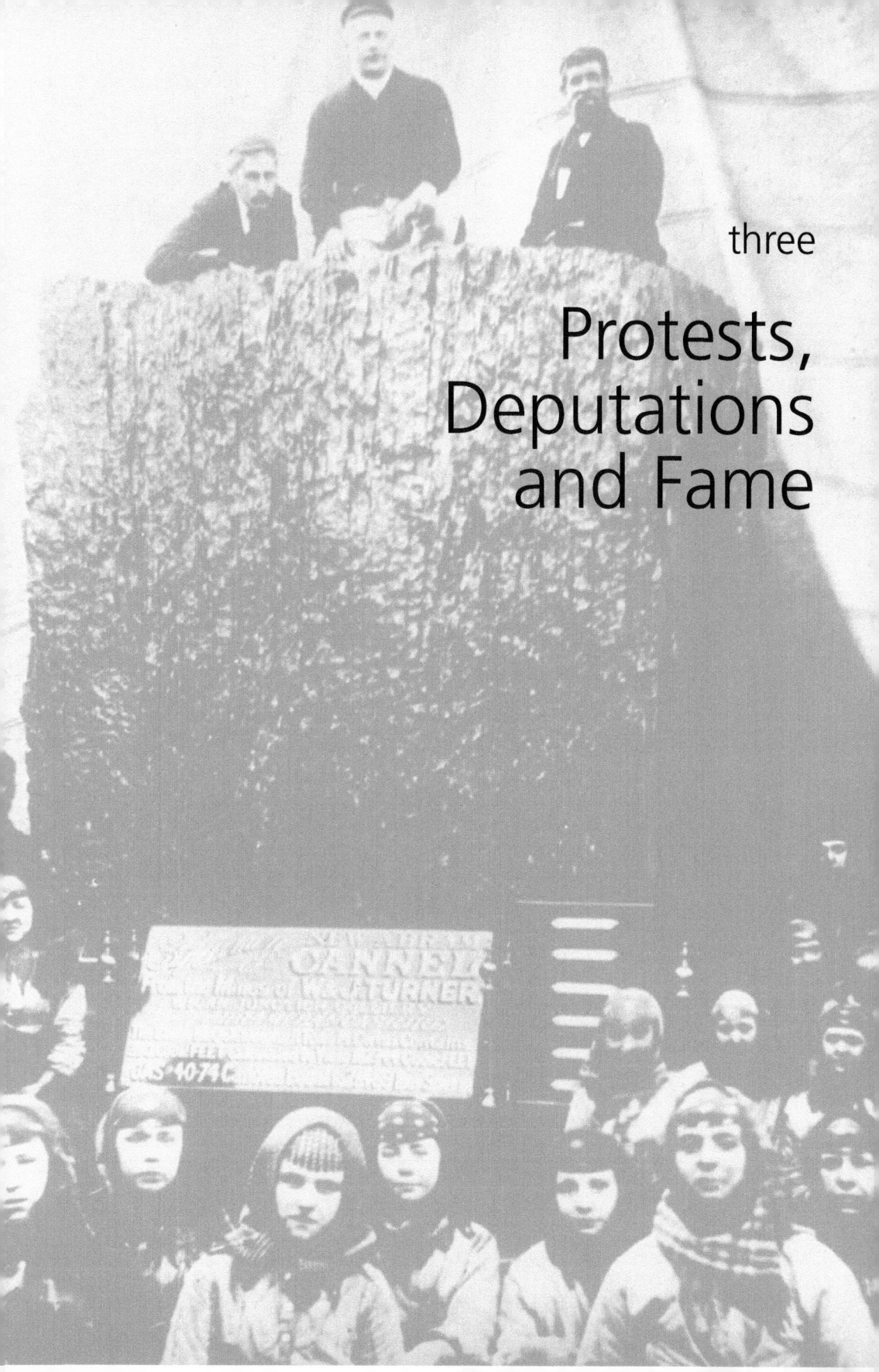

three

Protests, Deputations and Fame

In general from the late 1850s onwards women began to gain a significant political role and their situation slowly improved. Attention was turned towards pit brow women in 1859 when the secretary of the Miners Amalgamated Council appealed for information on grievances at work and suggested female pit work as one area of concern.

At the Conference of the National Association of Coal Lime and Ironstone Miners of Great Britain held in Leeds in 1863 these grievances were discussed and women's pit work came under scrutiny, especially in the area of their costume, work and immorality. A petition was sent to Parliament signed by 14,000 miners, one complaint being that:

> The practice of employing females on or about the pit bank of mines and collieries is degrading to the sex, leads to gross immorality and stands as a foul blot on the civilisation and humanity of the kingdom.

This slur on pit women was countered by Mr Gilroy, manager of Ince Hall Colliery, who produced studio photographs by Cooper of his pit women in their Sunday best looking as respectable as any woman in the neighbourhood. He mentioned that many of his pit women and girls regularly attended Sunday school. A further attempt to exclude women and girls from surface work in 1866 failed after a protest by the women themselves. Various articles in the press around this period criticised the women's work wear, *The Daily News* commenting in 1868 that, 'It is acknowledged that the habitual wearing of the costume tends to destroy all sense of decency'.

By 1869 female ratepayers could vote in borough elections, the National Society for Women's Suffrage being active from this year. In 1870 the Married Women's Property Act gave wives legal possession of any money they earned. No doubt the pit brow women aimed for this as much as possible anyway, to avoid the local pub swallowing up the family's earnings! From 1870 they could also sit on school boards and by 1875 were allowed to become Poor Law Guardians. The highlighting of the world of the pit brow women took a leap forward when Manchester-born Frances Hodgson Burnett, author of the children's stories *The Secret Garden* and *Little Lord Fauntleroy*, wrote her first play, *That Lass O' Lowrie's*, in 1877. Written in dialect, this was based on a fictitious town named 'Riggin' and there are no prizes for guessing where that was meant to be! High-profile accounts, even fictitious ones such as this, did not help the women's cause – quite the opposite! By the mid-1880s women pit brow workers were struggling to defend their legal right to carry out their work in the face of threatened legislation to exclude them. The campaigns to exclude women often came from those with a very poor knowledge of colliery work and particularly of the lives of women in mining districts and their unique sense of community. In 1886 Gladstone's Liberal Government allowed liberals the chance to argue for women's exclusion at collieries.

The major threat to the employment of women at collieries came in early 1886 with a new Mines Regulation Bill on the agenda including amendments to the Mines Regulation Act 1872. The clause to totally exclude females from work at collieries led to an eruption in popular support for the women, 1,407 of whom were at work at the Wigan area mines. Support for the women and girls came from all corners of the country. The local press in the Wigan area avidly documented progress on both sides of the arguments in great detail and surprisingly the wider national press also took notice of events and publicised them. We find rural vicars writing to *The Times* indignantly defending the women's right to their 'honest toil', men who had probably never even seen a colliery!

An example of the wider press coverage can be seen in this article from the *News of the World*, 18 April 1886:

The successful deputations in 1886 by pit brow women and their local politician supporters to the Home Secretary halted potential changes in mining legislation which might have removed them from their work. In 1886 Wigan mine owners who had mainly opposed the changes celebrated by commissioning Manchester artist Arthur Wasse to paint the women at work, the result to be displayed at Wigan Mining College. The pit head scene probably depicts Worsley Mesnes Colliery. The women are working in pairs pushing full tubs of coal away from the shaft towards the tipplers. Here the tub was turned upside down, its contents dropping onto screens below. The women then greased the tubs axles and sent it back down the pit on the next winding. The painting was exhibited at the Royal Academy in 1887, Chicago in 1936 and Paris in 1955 and today resides at Wigan College, Parsons Walk.

Meeting of Lancashire Pit Girls

A meeting of Pit Brow women, called by Revd Geo Fox at the request of the women themselves, was held in St. Peter's Church schools, Bryn, near Wigan, on Saturday afternoon. The chair was occupied by the Rev W J Melville, rector of Ashton, and there was a large attendance, the audience including about 200 of the pit girls employed in the district. The proceedings were most enthusiastic, and a strong determination was shown to resist to the utmost any legislation brought forward with the intention of prohibiting the continued employment of women on the pit brow. Mr Fox read letters from noblemen, clergymen, and others approving the agitation set on foot in the interests of the pit girls.

Left: *From the 1850s onwards the popular illustrated newspapers such as* The Graphic, The Black and White Budget, The Pictorial World *and* Illustrated London News *regularly featured engravings of colliery scenes both above and below ground. Here in the* Illustrated London News *in 1878, Wigan women are seen levelling off coal which has come down from the tipplers. Smaller sized coal then dropped through the screen bars into wagons waiting below.*

Below: *Douglas Bank Colliery, Wigan, around 1890, a photograph by Revd William Wickham (1849–1929) vicar of St Andrew's church. A keen amateur photographer, he documented life in his parish including work below ground and also the pit brow women. The colliery closed in 1920.*

Letter from Lord Crawford of Haigh Hall, Wigan:

The form of labour is without doubt severe, but it is a healthy one, as may well be seen by the bright looks of our pit girls when compared with the appearance of those engaged in many other trades. I cannot but feel that the wages thus earned are of material importance to numberless families in our district, and if stopped would in many cases bring about great distress through the enforced idleness of those who still have the right to be fed. In course of time many of the younger girls would be able to learn other means of wage earning, but that would mean a weary struggle, in which many would sink. So much for the physical points.

I do not consider that the employment of women at the pit brow is more unfavourable or more likely to produce moral injury than scores of other employment's; nay, I believe that were it possible to make an impartial enquiry into the question our pit girls would stay well to the front in quiet, decent, and orderly lives. As to the question of the clothing worn by them - the inheritance of their mothers and grandmothers - it may be dismissed without scruple as the senseless clamour of ignorant prudes, who, if left alone would probably put a 'frill' round the ankles of their kitchen table.

To sum up, I think that restrictive legislation would be: 1. Impolite, as reducing the number of bread-winners in the State. 2. Unjust, as forbidding the means of gaining an honest livelihood. 3. Cruel, as producing enforced idleness, more especially at the present moment, a time of great commercial depression.
Yours Faithfully Crawford.

Another unidentified writer was annoyed at outsiders voicing their opinions:

I have no sympathy with the occasional clap-trap which finds its way into the press about the immorality of our pit brow girls, and their unsuitability to make wives and mothers. This kind of trash evidently emanates from persons who have no knowledge of the work. However, there was much room for improvement in the work itself and in the surroundings ......
Yours etc

The Rector of Wigan (the Hon. & Revd Canon Bridgeman) said he did not think legislation on the subject was called for, but he should be glad if the attention drawn to it should be the means of inducing the coal proprietors and managers to make stricter rules and better separate accommodation for the pit girls during the dinner hours.

Canon Fergie of Ince wrote that he had no hesitation in condemning the present attempt to interfere with this form of female industry as absurd, unreasonable, selfish and unjust. Mrs Park, wife of the mayor of Wigan, and Revd H. Mitchell, vicar of Pemberton, wrote in the same strain. Addresses having been delivered by several speakers, a resolution was unanimously passed authorising a petition to be sent to the House of Commons against the proposed legislation.

The following extract from the *Manchester Guardian* of 5 May 1886 describes a meeting held at St Helens Town Hall but attended by Revd Mitchell of Pemberton. Here just about anyone, except for the women themselves, had a say. The journalist wryly notes this at the end!

Pit Brow Women

A meeting on this subject was held last night at St Helens, in the Town Hall. It was called by the Mayor, who had been formally asked to give the inhabitants the opportunity of taking action in a matter which closely concerns them. 'For the purpose of considering the intended interference, by Act of Parliament, with the employment of women on pit-banks, and the desirability of adopting a petition to Parliament in favour of securing the continuance of such employment' - these were the terms of the requisition, and as a demonstration in favour of the women the meeting was a great success.

The Rev H Mitchell, vicar of Pemberton, Wigan was the chief speaker of the meeting. The strong way in which he has elsewhere taken their part seemed to be known to the women, who gave him a pleasant reception. The hard things that have been said in various quarters appeared to take the women by surprise, but they readily adopted the view put forward by Mr Mitchell that outside their own districts there was dense ignorance as to the ways of pit

Pit brow girls operating the tub tippler at Douglas Bank Colliery coal wharf on the Leeds & Liverpool Canal around 1891. The tub probably held just short of half a ton of coal.

women. The attack on them, Mr Mitchell said, was only the first instalment of a crusade against all field work of women. There was to be no more harvest work, no more hop-picking, and very likely no more milkmaids. The women laughed heartily at this picture of the future. All through the meeting, indeed, they gave no indication of a feeling that their calling was seriously in danger. Arguments in favour of their right to earn a living at the pit bank they applauded with a heartiness that would have filled with envy a body of claquers, but remarks about want of employment and daily bread in danger seemed not in the least to depress them. Mr Mitchell produced some interesting figures. In his own Parish of Pemberton there are 160 women at pit bank work, of whom 133 are single women. If this proportion between married and unmarried women holds good in the rest of Lancashire, he estimates there are, in round numbers, 300 married women to 1,300 unmarried and widows, and of the latter number at least 1,000 can be classed as young unmarried women.

Were they, he asked, to be reduced to the miserable condition of the unemployed, or forced to compete with the already half-starved girls of Manchester with their 6s a week? If Parliament took away their right to earn an honest living on the pit bank it would be committing an act of tyranny which would disgrace the nation. Mr Mitchell is confident that 99 of every 100 Lancashire men would be opposed to such a gross injustice and he thinks there would be no fear of it if the facts were known. The Home Secretary has refused to receive a deputation of pit women, but Mr Mitchell believes is because he has already made up his mind that no Parliamentary action shall be taken against them. A resolution was proposed by Mr Councillor Forster that the Mayor be requested to prepare and send on behalf of the meeting a Petition to Parliament against the passing of any bill for the prohibition of the employment of women on the pit brow. The seconder of the resolution, the Rev A.A.Nunn, expressed the opinion that working at the pit brow was far healthier than working in the glass manufactories of St Helens.

As to the dress worn by the women, he said it was not only proper but necessary. It might be distressing, he added, to see women coming home from work with black faces, but comparing their moral state with that of other working women, he had no hesitation in saying that if their faces were black, their hearts were clean. A couple of miners afterwards addressed the meeting, but contributed no new facts to the discussion. The resolution was carried with great enthusiasm. It would have been interesting to hear the views of some of the women. The intelligence and frankness they displayed all through the meeting made it evident that the case would not have suffered in their own hands.

The *Wigan Observer* ran regular special reports on the controversial proposals from February 1886 onwards, including regional and national press features and letters, meetings of the British Women's Temperance Association and pit women's meetings held at schools and church halls.

On 17 May 1887 a deputation of twenty-three pit women, some taking along clean examples of their pit clothes, headed off to London to lobby the Home Secretary Mr Childers. The deputation of sixteen was made up of pit women from Bryn, Aspull, Pemberton, Garswood, St Helens, Haydock and Whitehaven. Accompanying them were former Pemberton St Johns church vicar, Revd Mitchell, the Lady Mayoress of Wigan, Mrs Park, and the wife of Atherton Collieries part owner, Mrs Burrows. This successful visit was widely reported, and the proposed legislation was withdrawn when the amendment came before Parliament on 23 June.

The *Leigh Journal* reported the victory on 20 May and at the end of the piece commented in a manner which typically summed up certain attitudes of the day:

Left: *Two pit brow women pose for Revd Wickham at Douglas Bank Colliery, Wigan, around 1891. The women probably lived in Holt Street off Woodhouse Lane which can be seen in the background.*

Below: *A rare and possibly unique image by Revd Wickham of pit brow women queuing on pay day at Douglas Bank Colliery. Dating to around 1891, the women's terraced homes in Holt Street are seen behind.*

Opposite below: Illustrated London News, *28 May 1887,* Lancashire Pit-Brow Women. *An engraving from a photograph by Wragg of Wigan. Some Wigan area women wore their working outfits on the deputation to the Home Office at the time of this image being published so this may be from a photograph of them.*

Above: Work Girls At The Wigan Collieries *as depicted in Casell's* History of England *around 1900. A stylised view with unusable pit headgears behind!*

THE COLLIERY "PIT-BROW" WOMEN.

The sympathy of sensible people, not unmixed with amusement, wonder, and a trifle of admiration, has been recently expressed in favour of the sturdy Lancashire lasses, who came up to London and called upon the Home Secretary, asking Government to resist the proposed clauses of the Mines Regulation Bill. Women and girls never go down into "the pit" at a colliery, but they are employed at the "pit-brow," the mound of earth and shale around the mouth of the shaft, in the work of screening or sifting the smaller coal. It is rough and hard work, though not more so than the labours of peasant women in the fields. It is certainly not unhealthy, for the strong creatures they are; and, as they are not associated with the men in this employment, there is no intercourse that is likely to prejudice the cause of morality. We can remember, however, some thirty years ago, seeing what was not a pretty sight: young women harnessed by leather belts to small carts, which they dragged up and down a line of planks, as if they were donkeys. The loads were not too heavy, and they seemed to think it no degradation. It is to be hoped, nevertheless, that this clumsy and unseemly method of carriage has long been disused. At any rate, the mere wearing of a "divided skirt" or loose trousers, of black flannel like the jacket, which forms a convenient and very decent costume, ought not to offend our delicacy, and cannot be deemed to justify passing a law to forbid their honestly earning their bread. A twelvemonth ago we received from one of the local clergy, the Rev. Harry Mitchell, Vicar of Pemberton, near Wigan, several communications upon this subject; and photographs were at the same time supplied by Mr. Wragg, of Wigan, some of which are now reproduced in our Illustrations. We believe the women habitually employed in this kind of work in Lancashire to be as good, modest, industrious, and well-behaved as any in England; and they are well able to take care of themselves.

But what are we coming to when Lancashire collier lasses storm St Stephen's and the Home Office and in homely Lancashire speech talk to the high and mighty ones who sit on the Treasury benches!

The victory of 1887 was widely reported leading to a wide range of coverage in specialist publications as well as the general press. These varied from *The Colliery Manager* to *The Lancet*, *The Illustrated London News*, *The Graphic*, *Pictorial World* and *The Leisure Hour*, extracts from which follow:

Colliery Pit-Women and Girls

Originally printed in *Leisure Hour* in September 1887:

The recent visit of pit-girls and women to London will be fresh in the memory of our readers. They came to protest against the proposed legislation for abolishing female labour in connection with collieries, and they were successful in their mission. The deputation appeared in costume, and won general approval by its modest yet independent bearing. It was supposed by some well meaning philanthropists that all colliery labour is degrading to the sex, and that only men and lads should be employed in work so hard and apparently so repulsive. But the work of these women is not the pit's mouth, and therefore in the open air, under easy inspection, and it is apparently far less injurious to health than factory labour, and not more laborious than agricultural labour, in which many women are engaged. These may be seen any day at the town of Wigan. Watch them as they pull the trucks of the precious mineral from off the cage as it clatters out of the dark abyss. The collier-girl has a knack of doing this dextrously. She and her fellow workers reload the cage with empty – 'boxes,' then run the full ones to the edge of the stage upon which the winding-gear is embanked, where, by means of a lever, the coal is tipped over on to a long sloping wooden channel called 'a shoot.' Other girls are occupied on the shoots; these shovel the lumps forward with huge spades, pick out the bits of stone and slates, and see that the finer dust (used for making coke) falls through the 'screen' a kind of riddle formed of iron bars. Others again are filling the coal from the stacks into wagons that stand at the railway siding close by, and at times they will climb into the wagon to heave about or straighten the load. The dress of these Wigan ladies is peculiar. Their nether garments are not very womanly. A few wear men's coats and waistcoats as well as breeches, so that often as we have walked behind these when slouching from their work along the streets, we have determined their sex only by their earrings, as well as perhaps by the bunch of hair done up in a kerchief, and half drooping from under an ordinary mans cap. A stranger is rather startled and perhaps feels some revulsion upon beholding these Amazons at the pit brow yet he should not hastily condemn their employment as degrading. The girls like the work, and appear to be very healthy. An argument in favour of the occupation takes the following form: the widows of miners and unmarried women in exclusively mining districts have no other means of getting employment; numbers of widows (of miners or otherwise with families thus earn a satisfactory wage. For orderly behaviour these workers are said to compare favourably with mill operatives. As a rule, colliery girls are married (often at an extremely youthful age) to miners; still their occupation does not preclude them from being chosen by other workmen, as iron-foundrymen, spinners, masons etc. As for their dress, it has been begotten by custom as well as convenience, and, considering the nature of the work it is both decent and becoming. Many are at work together in their variously coloured head coverings, we see something quite picturesque in the neat and rapid precision with which they draw the filled trucks off the cage, hurl in the place of these the 'empties' to go below to the far-off coal getter, and then troll off their burdens to empty them over the embankment. The agility of every movement seems the more noticeable when it is recollected that each truck (itself weighty and bolted with iron) contains a load varying from six to seven and a half cwt. It may be added that the girls have to work from six in the morning to four in the afternoon, with an intermission of half an hour for breakfast and an hour for dinner.

A man whose congregation at this time was virtually made up of miners and their surface working wives was the Reverend William Wickham (1849–1920). He had arrived as curate of St Andrews Church of England church off Woodhouse Lane in 1878. Mining districts were not new to him though, his first appointment being at Talke-o'-the-Hill in Staffordshire where the local mine had suffered a massive explosion in 1866 killing over ninety men and boys. A keen amateur photographer, he documented many aspects of the life of his parish, even to the extent of taking some of the first ever photographs below ground in local coal mines. Initially holding reservations when offered the position at Wigan due to its reputation, he nonetheless immersed himself in Wigan life even to the point of wearing clogs. He was, however, as he admits, beaten by the local dialect which he struggled with for many years.

He took approximately 200 photographs in the Wigan area, including many of the local colliery workforce, above and below ground. His informal images of women working on the surface in 1891 and during the long mining strike of 1893 are today of enormous importance, on a national level also. Knowing the photographer as they did, the subjects are nearly always relaxed and natural in pose. Wickham gave regular slide shows of his photographs in 1891 and 1893. These were very well attended by the local gentry, miners and their wives and caused a great deal of amusement, in the same way that the recently discovered motion films taken by Blackburn cinematographers Mitchell and Kenyon were to do from 1897 onwards. At the time of the production of Revd Wickham's images, three other photographers with studios also produced images of the Wigan area women surface workers: John Cooper, Rudolph Douglas, James Millard and Herbert Wragg. Examples of their work can be seen at the end of this chapter.

Around this period a number of individuals began to describe their former colliery work in writing once they realised the general public had a strong interest. A Mrs P. Holden looked back to working from 1890 onwards at the collieries around Chorley, Duxbury and Adlington. Her very detailed account of the brow woman's actual job is well worth including. Even though she worked just over the borders of present-day Wigan Borough, Wigan area women and girls would be working at the pit she mentions. Mrs Holden walked three miles to work, and some others walked four or five. Any terms needing clarification are in brackets.

At the age of 13 I started to work at a coal pit, called Duxbury Park Colliery, near Adlington. I had 3 miles every morning to go to my work, so it tied me to get up every morning at 4 o'clock, as I did odd jobs about the house, before setting out to my work, at 5 o'clock. So off I set out, with my basket on my arm, and a full can of tea in my hand. I wore a red head wrap, tied around my head, to keep the coal dust out of my hair, then a nice shoulder shawl thrown over my head wrap. I wore a black velvet blouse, and a blue striped pit skirt. I made my own pit brats, out of Irish linen. I wore a man's jacket to come home in, also pit breeches as well, I took a pride in my clogs, they shone like a raven. Off I set out to my work, singing to myself, happy as a lark.

We used to wait on brow for full tubs of coal coming up in cage. The brow man would be waiting for the cage coming up, he would lift up the lever, and push the full tubs out of the cage with his foot. No time lost, the lasses grabs the tub, and runs it to a shoot, which has a tippler attached to the shoot. We kick the tub of coal down the shoot and it runs into a wagon, in rail road below. Now that shoot is called through and through, because more dirt than coal goes down it, copperas [brasses or Iron Pyrites] as well.

Now that coal is to be stacked in rooks [heaps], until we get enough empty wagons, to carry on our own work in the screens. Now, if you'll take notice, stacked coal, if stacked too long, it starts smouldering [spontaneous combustion]. That is with copperas, being left in the coal, when it is being stacked. Now there was three shoots, and two steel moving belts, each shoot had a shaker attached to it, an iron riddle, as made slack. There was a cobble belt, nut belt, there was seven lasses on a belt, three on each side, then one lass close to shoot, with an iron rake, ready to spread the coal out, as it was tippled down the shoot.

On the belt, at the other end of the belt, there was a very small shoot, as the coal left the belt it dropped down the small shoot, into a eight ton wagon below screens, in rail road. If it was blowing or raining, the coal dust would fly up small shoot, into your face and eyes. We was sending as many as twenty or over wagons of coal out per day [ten ton capacity]. Now we had to pick copperas and dirt out of the coal. Now the copperas was thrown into a copperas box until we had picked ten ton, then after it had to be thrown on the belt and run into a ten ton wagon. That was sent away to big works to be melted down for different purposes [one of which was the production of sulphuric acid].

Now the dirt as we picked out of the coal during the day had to be thrown at back of us until we knocked out at 4 o'clock. Then the brakes man would shunt a dirt wagon under the belt so we had to start again shovelling it on belt, and we was dead tired. It took us an hour to fill a dirt wagon. We was leaving pit at 5 o'clock, and I had three miles to walk back to whoam [home].

Next day, we would be on a sample wagon, a big order had come in for House Coal Nuts, and not until that wagon was full dare we look up. While picking out dirt copperas and even shale, our fingers bled at the end. We had to do it, forsake of us losing fresh orders. We all had an iron chipper, that's to chip copperas off coal, and throw remainder on belt, and pick out the dirt at the same time off belt, as was going past us. I worked six days per week and never once late. As I had to iron my own clogs, I put two sets of irons on my clogs, the small set in the middle. They were pretty heavy, but I never bothered if I was saving mother something.

[Describing trimming or levelling off the ten ton main line wagons] The first wagon I got on the wagon bumper [buffer], then into the wagon; as the wagons kept moving down the lines, as I was trimming them making room for other empty wagons, to be shunted, under the screens or shunts. By the time I had trimmed my wagons, it had got dinner time. There was an iron door across the shoot [for filling coal into the wagons] with iron lever attached to the iron door. Now that was to stop the force of the coal for going too swiftly into the bone coal wagon, below in the rail road below the shoot. Every now and again, the brakes man would lower the bone-coal wagon as it was getting full. Now the coal in the shoot would be held up until the brakes man would shout, 'Kech-um-oer'. Then I would be quick, lift up the lever, while about two tons of coal slided in the wagon below.

Now, as that wagon was being filled, the pit engine would draw along side of my wagon. I had to leave the iron-lever, jump into the bone coal wagon while in motion. I would have to hand large cobs of coal to the engine driver, until his coal box was full. By that time, my wagon had got full of bone coal, the pit engine would draw the wagon from under the shoot down the rail road as I had trimmed, the same wagon as it was being filled. We was sending as many as twenty or over of bone coal out each day. It was steam coal, and, we was in need of it for our ships as there was a war on, South African War [Boer War 1899–1902]. We could not afford to lose a second, and the miners, brow men, lassies, worked very hard.

Now if we wanted to have a drink of water, we had to go to the Throstle Nest Wood, fill our cans under the spring as the water at the pit tasted of oil. Now the lasses got a bit fed up working on owd brow [the old pit brow nearby] as there was a furnace [for ventilation] down owd pit, and the sulphur as came up gave us all sore lips [effectively weak sulphuric acid vapour!] and bad colds and we all kept sneezing.

We started working in full swing until about 1905, there was a strike among miners. They got to know as our pit was in full swing, and the strikers got to know as our pit was working. Morning after they landed, with all sorts of weapons, some had pick arms. One or two of the mob, went into winching engine house and stopped him for winding coal. Three or four, went on brow and, as the cage came up pit with full tubs of coal, the strikers said, 'Think on, next tubs as come up the shaft, will go down, and not in the cage either, un thee on top of it'. So the browman had to walk off brow, and the mob took command, while all the miners came up pit. Most of them, had brought their jack bit [food] with them from Wigan, so we all knocked out, and went whoam, in our black faces. There was only odd men [below ground] working on water jobs, keeping water out of mine places, at far end.

I have often cried, when I've seen the miners be brought up pit, dead or wounded, wondering, who is he, or were he came from. Sometime, he has been under a fall, I mean, stone, and it takes longer to get him out. The head gears is going round very slowly, and everything goes silent. I think a miner should be treated in a proper way, the way he is to risk his life, besides having lumps knocked off him, while he is digging for coal, for you to burn. This is my point of view:- Treat the miners in the right way, then in return he will treat you, by getting out the coal, and if you will sit back and think, the miner is the Back Bone of England.

The news of the 1887 victory also spread around Europe and was particularly noted in countries such as Germany, France, Holland and Belgium where women worked on the surface. Curious French visitors came to Aspull in 1902 and recorded their findings at Wigan Coal and Iron's Crawford Pit as follows:

If we visit the important mines of Wigan, we see female workers occupied on the pit site as in France. Certain ones receive at the pit bank, at the entrance to the shaft, the tubs full of coal which they push towards the mechanical tippler placed above the screens and picking belt. Others are employed in sorting the coal; they form a complete group which is a curious sight by the type of costume and the facial expressions, some of these people being small girls rather than teenagers. The frail limbs and childlike faces of some of them indicate that only light work could humanely be asked of them, so that concentration plays a greater part in their work than muscular effort. Their costume, in other ways very simple, consists of a type of trouser which only just covers the knee, and which has a very short skirt over it which is almost hidden by an apron whose light colour, sometimes white, makes a singular contrast in all the sombreness; they wear over the top half a Scottish shawl of various colours and design. Finally, they wear on their heads a type of hat which protects them from the coal dust. It must be added that they wear on their feet a type of boot, the upper part being in leather, but whose sole being extremely thick is often made of wood. The picture of different groups of sorters shows the exactitude of what was stated earlier, that is to say, that these young girls have a taste for feminine things and a love of ribbons. Most of them in fact wear around their neck ties, whose folds will soon become nothing more than little nests of coal dust.

It is interesting to note that often descriptions of pit brow women centred on mannerisms, clothing and the harshness of the job. The women were described as though they were unique beings, devoid of gender. The authors often appeared to be blind to the fact that these were in fact just ordinary hardworking women who were prepared to get stuck in and get their hands dirty, a not uncommon sight in northern working class districts!

In 1891 what was to be an influential local publication appeared, *Sketches of Real Life at our Collieries* by Samuel Woods. Samuel's thinking was a strange mix and came from all directions and influences, being as he was a Conservative, a coal owner's son, yet also the Miner's Agent based at Park Lane Colliery, Garswood, and from 1907–11 he was the town's mayor. Samuel had, under a nom de plume, submitted various articles on local colliery life to the Wigan Observer. The recent high profile given to the pit brow lasses meant that Starr, printer and bookseller of Wallgate, was keen to publish a compilation of Sam's articles. It is worth noting that even in the midst of one of the busiest coal mining districts in the world Sam thought that some local residents whose knowledge of mining life was misinformed through 'tainted public journals' were in need of re-educating as regards the communities alongside them!

In his piece entitled *The Pit Brow Girls*, Sam created a fictitious character, miner's daughter Jane Armstrong, an intelligent girl who had aspirations far beyond working on the pit brow, yet who lived amongst girls who worked there. Her friends try to convince her that work on the brow is good honest respectable toil, the few shillings paid being vital to help the family maintain its respectability. Jane is eventually won over and like many other girls forgets her ambitions. Sam's own ideas on working class society and inequality were acquired in rather an obvious way!

Jane argues to her friend Ellen:

> I don't quite follow you in your reasoning that some girls were born to be ladies and others to be the mere drudges in society… I believe I was designed to be a lady as much as our minister's daughters, or any other girl and I should have liked to have been one. I could have learned both Greek, Latin and French, as well as the first fashions of the day. But why are we not ladies Ellen? I will just tell you. It is because those who are now rich have always predominated, and have crushed the poor down. They have made their own laws, which are made in such a way that the poor will never be able to get on a level with them, and they will always have to be in subjection. I wonder why the poor people don't rise up in a body and demand justice! I may say Ellen, the last part of your statement has almost persuaded me, notwithstanding my strong objections to give the pit brow a trial… for I know that it is hard work for my father and my brother to maintain all of us and to this end, if you think I could get a place on the same brow as you work at, I will try and put aside my prejudice and begin working at once on the pit brow, like you and other girls in the neighbourhood.

Sam continues:

> On the next morning Jane Armstrong, fully attired in the coarse garb of a pit brow girl, in company with her friend and associate, wended her way to the pit brow and from that day to this has continued in the same employment. In Jane Armstrong's history we see fully illustrated the truth that 'the hard irony of fate' which surrounds the position of the thousands of poor girls in our rural districts acts as a retardant in stifling lofty desires and aspirations, and in preventing them from rising to higher and better positions in life.

In 1897 another local publication, *Mining in the Victorian Era*, by the eminent mining engineer and principal of Wigan School of Mines, C.M. Percy, describes cleaning the coal at the colliery surface. He mentions that coal washing, recently arrived, was very expensive and that few collieries had adopted it. Along with comments on the employment of women and girls on the brow he mentions as he terms it the 'New Woman', a short-lived phenomenon in his view:

> A good many well meaning citizens in Parliament and out of Parliament - especially those who know nothing of mining - would abolish all female employment at collieries. I am no admirer of the 'New Woman' who I think, although conspicuous for a short time has not come to stay... The work of cleaning the coal on the surface at our collieries is as suitable for women as any of these (*other female occupations*) and much more pleasant and healthful. If all men's work is to be done by men, and all women's work is to be done by women, a good many men will have to retire. We had better leave well alone.

In 1911 when the Coal Mines Regulation Bill was being read the question of allowing the continued employment of women and girls on the surface at mines was again raised and an amendment to exclude them proposed. This was during a period of high unemployment; the miners were asking for a minimum wage and the suffragette movement was gaining strength.

By late July *the Wigan Observer* was monitoring the course of the popular debate. Mass meetings in support of the women took place in Wigan in August, the Wigan Mayor Sam Woods (in office 1907–11) and Stephen Walsh MP speaking to the crowds. On Thursday 8 August 1911, a protest delegation accompanied by the Lady Mayoress of Wigan, Mrs Woods, headed off to London, this time to the House of Commons itself. This included forty-seven pit brow women from Abram, Coppull, Haydock, Hindley, Ince, Pemberton, West Leigh and Westhoughton. The girls created quite a sensation as they headed through London towards Whitehall in their work wear and clogs or 'wooden boots' as they were described in the press. The following day was a rare and much appreciated opportunity for a group of Lancashire pit brow girls to see the sights of London.

On their return *the Wigan Observer* kept the public interest alive with features by Sam Woods in August and others (including photographs) on the women and their work in October. Interestingly in October the paper featured a meeting of the Women's Social and Political Union 21.10.1911 with local MPs Stephen Walsh and Mr Neville speaking. In late November, as decision time on the amendment approached, Sam Woods lobbied certain wavering MPs to ensure a successful outcome for the women.

147, Wigan Lane,
Wigan,
November 20th, 1911.

Dear Sir,

With reference to the Coal Mines Bill which will shortly be considered on Report Stage by the House of Commons, whilst the Bill was before Committee the following amendment was carried:

'No girl or woman other than those employed on or before the first day of January 1911 shall be permitted to be employed above ground on any Mine, provided always that this section shall not apply to any woman who is engaged in the cleaning of Colliery Offices or for any other like purpose.'

On behalf of the Pit Brow Women I appeal for your vote and assistance to defeat the amendment when the same comes before the House. Over 5,000 women are engaged in the work which has been proved to be suitable for them, to be neither detrimental to their health or morals nor dangerous to their limbs. The women like the work and bitterly resent this attack on their livelihood. As a class they compare favourable with any other class of working women, and it will be a great injustice to the women of the country if Parliament debars them from engaging in this work.

May I ask your careful perusal of the enclosed:

A DESCRIPTION OF THE PIT BROW GIRLS' WORK, and statements made by;

Dr. T.M. ANGRION, of Wigan, who has a large Colliery Work-people's Club Practice, extending over twenty-five years;

Dr. COOKE, of Aspull, who has had over thirty years' experience in a purely Colliery District;

Miss MACINTYRE, Matron of the Royal Albert Edward Infirmary, Wigan, which receives Indoor and Outdoor Patents from a radius of seven miles of Wigan, in which area over 2,000 Pit Brow Women are engaged; and

The Rev. T. F. B. TWEMLOW, Vicar of St. Peter's, Preston, and for several years Vicar of Abram, in which Township several important Collieries are situated, and where many women are employed.

I hope these statements will interest you and enable you to decide to support the Pit Brow Women's case by your vote and influence.

I have the honour to be,

Your obedient servant,

Sam Wood
Ex-Mayor.

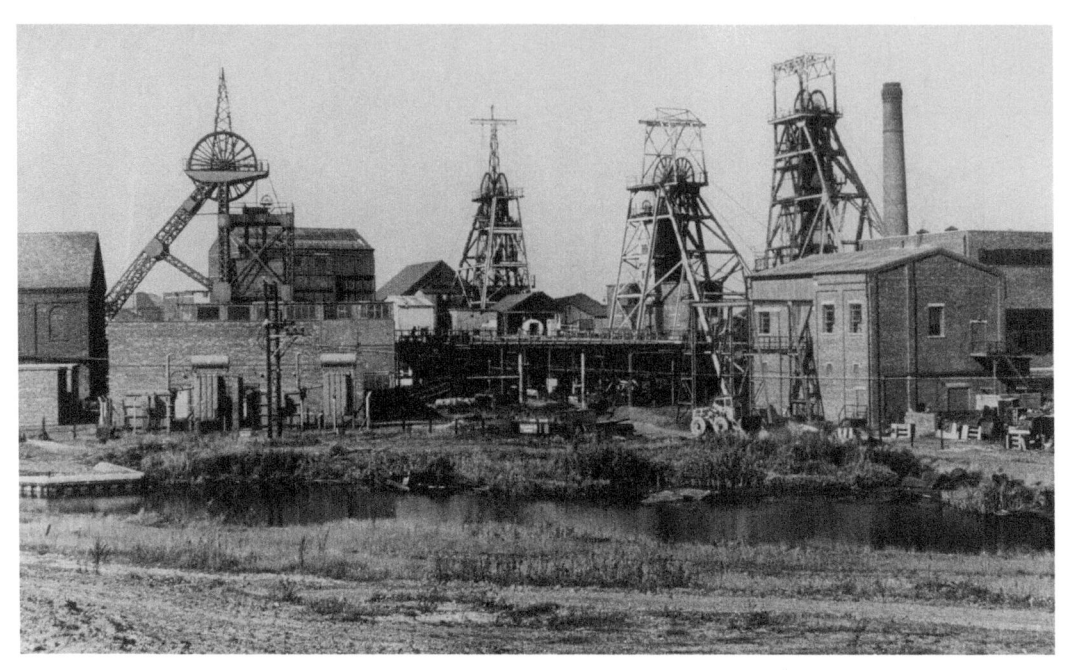

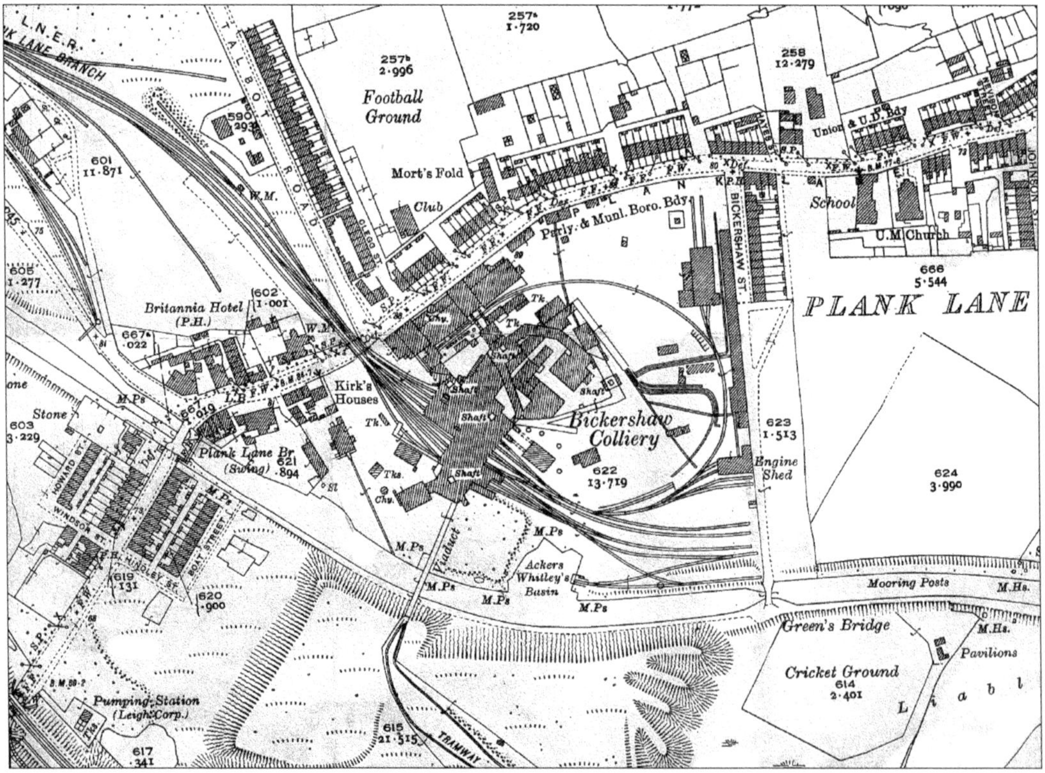
L.N.E.R.
TALBOT ROAD
257b
2·996
257a
1·720
Football
Ground
258
12·279
Union & U.D. Bdy
590
293
601
11·871
W.M.
Mort's Fold
Club
Parly. & Munl. Boro. Bdy.
BICKERSHAW ST.
School
U.M.Church
666
5·544
PLANK LANE
605
1·277
602
1·001
Britannia Hotel
(P.H.)
667a
·022
Tk
Chy.
Shaft
Kirk's
Houses
Stone
603
3·229
667
·019
Plank Lane Br
(Swing)
621
·894
Bickershaw
Colliery
622
13·719
623
1·513
Engine
Shed
624
3·990
HOWARD ST.
WINDSOR ST.
HINDLEY ST.
BOAT STREET
619
·131
620
·900
M.Ps
Viaduct
Ackers
Whitley's
Basin
Mooring Posts
M.Hs.
Green's Bridge
Pavilions
Cricket Ground
614
2·401
Pumping Station
(Leigh Corp.)
617
·341
615
21·315
TRAMWAY

Above: *Photograph by George Knight, a Leigh jeweller of Bickershaw Colliery pit brow women in 1886 during the debate about their employment. The giant of a surface foreman is Thomas Burns, an engineer who invented various safety devices for winding engines.*

Right: *A superb image taken in 1893 at Wigan Junction Colliery near Abram, showing possibly all of the colliery's pit women and girls. A distinctive three-plank main line wagon stands behind.*

Opposite above: *Bickershaw Colliery, Leigh, seen from the south beyond the Leeds & Liverpool Canal (Leigh Branch) in 1970. Sunk in the 1860s, the colliery employed women on the surface until the early 1950s. At its peak five shafts were in use. It closed in 1992.*

Opposite below: *Bickershaw Colliery, Plank Lane, Leigh 1928. Colliery waste was being tipped to the south where subsidence had created The Flash. With five shafts in use houses, schools, churches, pubs and clubs surround one of the largest collieries in Britain at that time.*

Left: *A colliery scene including women pushing tubs, probably at Moss Pit, Ince, by Taylors of Platt Bridge around 1905. The view gives an idea of the dimensions of a typical wooden pit headgear of the time.*

Below: *Twenty pit brow women and girls, possibly at Westleigh around 1900. The tartan head scarves with fringes are typical of girls at the Leigh and Atherton collieries.*

Opposite above: *Brow women and girls at Meadow pit, Aspull, north of Wigan around 1895; William Pit is in the distance. A French mining engineer visiting Aspull in 1901 stated, 'The frail limbs and childlike faces of some of them indicate that only light work could humanely be asked of them, so that concentration plays a greater part in their work than muscular effort'. I wonder what the women and girls would have thought of that comment!*

Opposite below: *A fascinating but sadly unidentified surface group at a Wigan area colliery around 1900. Here possibly all the forty-four surface workers are shown, illustrating a very wide age range, seemingly from about thirteen to seventy.*

Number two from the 'Pit Brow Girls' series of postcards, c.1905, published by Thomas Taylor of Platt Bridge and probably photographed by Starrs of Wigan.

A sparklingly clean pit woman wearing women's clasp clogs. No.5 of the 'Pit Brow Girls' series of postcards, c.1905, published by Thomas Taylor of Platt Bridge.

Girls screening coal.

Above: Girls Screening Coal, a postcard by Starrs of Wigan and published by Taylor of Platt Bridge around 1905. Probably taken in the screen sheds at Strangeways Hall Colliery, Hindley, where other pit women views were taken. The girl is using a long plain rake to spread out the coal, allowing small material to fall through to the belt below.

Right: Although a poor image, this view of 1893 is worth including due to the young pit girls in the foreground. The event is the raising from Wigan Junction Colliery near Abram of an 11-ton 14cwt block of the Wigan area speciality, cannel coal. This was intended for the International Chicago Exhibition.

Above: *Distinctive headgear worn by pit women at Gib Field Colliery, Atherton, around 1890. Working outfits varied amongst mining districts and even at adjacent pits.*

Left: The Daily Graphic, *21 September 1893. Striking miners and their families collecting coal off the waste heaps at Platt Lane near Whelley, Wigan. Women are in evidence and some will be pit brow workers well used to handling coal.*

Above: *The 1893 strike, a scene at Howe Bridge Colliery, Atherton, looking east. The company allowed their workforce to collect coal from the waste tips during the sixteen-week dispute. Inefficient screening meant that large amounts of coal could be found and the company's pit brow women will no doubt feature in this photograph.*

Right: *'Pit Brow Girls', card No. 8 of the Taylors' series of about 1905, possibly taken at Bamfurlong Colliery.*

Later that month, as a direct result of the various submissions mentioned, the amendment was defeated and The Coal Mines Act 1911 emerged. For both boys and girls, thirteen was to be the minimum age of employment. For boys, girls and women a maximum working week of fifty-four hours with a daily maximum of ten hours was to be enforced. Boys, girls and women working over five hours were to be allowed breaks totalling half an hour, and those working over eight hours breaks totalling one and a half hours.

The Act reassured those who had been vociferous in their opposition to boys girls and women working on the colliery surface, adding:

> No boy, girl, or woman shall be employed in moving railway wagons, or in lifting, carrying, or moving any thing so heavy as to be likely to cause injury to the boy, girl or woman.

In reality women at Maypole Colliery, Abram were still single-handedly stacking 7ft pit props in the yard ten years later. Ranging between 5 and 6in in diameter and occasionally larger, these props, especially when soaking wet, might weigh from 50kg (approx. 110lb) upwards.

Nellie Potter remembers that at Maypole meals were taken in a cabin with often company of a sort she would rather not be reminded of:

> One day there was a woman from Standish, picks her shawl off the hook, puts it on, she felt something running down here. Her screamed – a mouse. I used to be scared stiff of 'em coming in my basket.

Pemberton Colliery, Wigan, October 1931. From the left: Queen Pit, King Pit, Prince Pit and Old Bye Pit, an old colliery sunk around 1815. At the time of the photograph, 473 surface workers were employed, including many females, some of whom went on the deputation to the Home Office in 1886. The pit closed in November 1946.

The forty-nine pit brow women at Pemberton Colliery in 1911 at the time of the deputation by colliery women to London to safeguard their jobs. Some of these women formed part of the deputation.

A carte-de-visite of pit women at Pemberton Colliery, Wigan, by Millard of Wigan. The coal is edging along the 'shaker' mesh conveyor which oscillated to and fro, the girls spreading the coal out to allow small material to fall through. They also broke up very large lumps of coal into manageable sizes and chipped off any shale or stone attached.

Lamb and Moores Newtown Colliery, Wigan, around 1900. The colliery employed pit brow women extensively on its old fashioned screens and coal shutes.

Women at work on Lamb and Moores Newtown Colliery coal shutes. Coal from the screens of varying sizes or qualities came down the shutes to be bagged for local coal dealers, Thomas Hatton calling on the day of the photograph. Working in the open, these women would have had a particularly hard time in winter. Note the gas lamps.

PIT BROW GIRLS 1.

In the screen sheds at Ince Moss Colliery around 1910. At this pit it appears the fashion was for women to wear a long dust apron over their shifts.

Opposite above: *The first postcard of around 1905 in the Taylors' series 'Pit Brow Girls'. This has to be the classic pit brow women view with the tall pit headgears towering in the distance. Taken at Wigan Junction Colliery near Abram, which was sunk in 1887 and closed in 1962.*

Opposite below: *Pit brow women at Fletcher Burrows & Co.'s Gibfield Colliery, Atherton, pose for* The Daily Sketch *photographer on 22 August 1918. The paper ran an article showing how everyone was doing their bit during the war including pit women.* The Daily Sketch *was initially established in Manchester, later opening a London office.*

Above: Pit Brow Lasses At Work, c.1905, probably taken at Moss Pit Ince. At the rear stand the checkweighmen, who monitored the weight of coal sent up the pit by each man, the management checkweighman double-checking the weighing!

Right: *A studio image of pit brow girls working at Park Colliery, Garswood, around 1920.*

Right: *A superb Millard of Wigan studio carte-de-visite of around 1900. What a colourful scene this must have been, sadly recorded in shades of grey. The heating grill on the floor rather ruins the pretend landscape effect!*

Opposite below: *Pit brow women at J.&R. Stone's Park Colliery, Garswood, in 1900, a photograph taken for the Mines Inspector's Report. The colliery was sunk in 1887 and closed in June 1960.*

Douglas Bank Colliery Girls At Work, *a postcard of about 1905. Here the pit bank is covered over. Note the metal plates on the floor to allow tubs to be taken off the rails to be upturned and checked for their content.*

No.6 in the Taylors' 'Pit Brow Girls' series, taken at Moss Hall Colliery, Ince, Arley Pit, around 1905. Note the double tub hoist to the right.

The Pit Head, *a postcard of around 1905 probably by Taylors, of Strangeways Hall Colliery near Hindley Pit. Pit brow women stand on the raised 'heapstead', as it was known. As well as working at the screens they would have loaded up the pit props seen in the foreground which when soaked would be quite a weight.*

Strangeways Hall Colliery around 1905. Young and old females appear in this photograph taken from the same angle as the previous shot.

Pit brow women at Gibfield Colliery, Atherton, in 1905. The women are laughing because the second row is made up of men dressed up to fill out the numbers! This is just part of a very detailed image Fletcher Burrows & Co. had produced to both use in lantern slide shows and also produce postcards. Director Clement Fletcher

Colliery Lasses, a Wigan area postcard of around 1905. The women are wearing waterproofed aprons on top of their shifts and appear none too pleased to be photographed!

was very interested in the colliery's history and was very generous to his workers, who lived in cosy cottages and had leisure facilities, a school and a church, along with public baths.

On the shaker screens at Chanters Colliery, Atherton, in 1905. Living in a close-knit mining community, these women were probably all married to Chanters Colliery miners.

Girls Screening Coal, *No.2, a postcard of about 1905 published by Taylors of Platt Bridge. The women are wearing their husbands' old jackets to keep warm.*

A happy group of surface workers at Hindley Colliery around 1910.

All the family and no doubt women surface workers helping out coal picking at Pretoria Pit, Atherton, during the successful 1912 minimum wage strike. Two years earlier an explosion at the pit killed 344 men and boys, the third largest in British mining history.

Wigan Coal and Iron's Alexandra Pit, Whelley, around 1920, sunk in 1875 with shafts eventually 777 yards deep to the Arley seam, and which closed in 1955. A rare and early motion film of about 1912 produced by the company after the recent deputations to London featured the women at work on the surface.

A still image from the film of Alexandra and Lindsay pits around 1912 showing a fairly old pit brow woman covered in dust and desperately in need of dental care!

Seventy-three similarly attired pit brow women pose here at Pearson and Knowles Ince Moss Colliery around 1910. Behind them are the screening sheds and tub hoist.

Gibfield Colliery, Atherton, in August 1918.

Gibfield Colliery again in 1918, this view taken inside the screening sheds.

Twenty-four brow women pose beside the shaker screen at Gibfield Colliery in 1918.

four

Decline and Fall of the Pit Brow Lass

After the long period of opposition to female employment at collieries from the 1860s through to 1912 women and girls on the pit brow were finally left alone to subsidise their family earnings. The main threat for the future of pit women's employment and not of a political nature had arrived during this period in the form of coal washeries. One of the earliest installations nationally appeared at Gibfield Colliery, Atherton, in 1892. Washeries were expensive though and most collieries in the meantime carried on dry screening and employing women at the belts. A woman whose name is not recorded recalled working on the screens at Chanters Colliery, Atherton around 1910 where hand picking and coal washing were in place alongside each other for many years to come:

> I went on't pit brow at sixteen until my twenties. There were six belts, there were three that had cobs on and there were three that had what we would call 'burgy' a smaller coal. You started on Number One and you worked yourself up to Six, and then you were called as a helping hand, like if there was anyone off on one of the belts they would put you on. As the coal came down, if there was any dirt in it or anything, you used to have to lift these cobs of and put them on a table and you had a little pick and you used to have to chip all the dirt off, and then you threw all that coal back again on't belt and it used to go down into a wagon beneath.

Nellie Potter recalled her pre-1926 strike days on the brow at Maypole Colliery Abram in April 1984 for a feature in *The Observer Magazine*. She not only worked in the screening sheds but also worked on the pit bank pushing half-ton tubs around:

> I was at the Maypole Pit until the 1926 strike. The banksman pulled the tubs out of the cage, we pulled the tubs over the rails and then I had to pull them on to what we called creepers – two forks caught on to the axles – and up to the checkweighmen.

Also working at this period was Margaret Calland, born in 1900. She pushed tubs around after separating coal from dirt at the conveyor belts. She then pushed the tubs up to the weighmen:

> There were always two, a union man and bosses' man. There was a tally on the tub and I'd shout out the number so they knew which miners it was. This was the olden times. Many a time we'd throw out as much dirt as we could before it reached the weighmen. Usually the pitmen were as careful as could be, for the sake of not getting fined for too much dirt. And of course we knew all the pitmen, my father was one and my husband. I wore a short apron and a square shawl, you put paper with wire round the edge, round the front of your head to keep the dust off your hair, and each corner of the shawl was pinned to the side of your chest.

Nellie Potter worked at Maypole Colliery, Abram, a couple of years after the disaster of 1908 when seventy-five men and boys died after an explosion. Working on the screens in winter was very harsh:

> There wasn't a cover nor nothing. We didn't keep dry we was wet through and when it was winter and they were hard winters, they (your clothes) nearly froze to your back, it was terrible sometimes.

Asked about keeping her clothes clean:

> Oh you just dumped them at one side often than not there'd be three sets, one drying for another if it was wet, bad weather.

Serious accidents did occur to pit brow women and girls, the unfenced machinery around them being the main danger; Nellie remembers an accident at Maypole around 1919:

> I remember one girl... and she went under some belts. There was what we call a little cross belt and she went under this on to this belt and she got whirled around the shaft. She died, it killed her.

The arrival of the First World War saw an increase in the numbers of women employed at collieries, with up to 11,300 working in such a way nationally by 1918. The taking on by women of previously unavailable roles in industry influenced the development of women's rights in general. Women around Wigan eagerly showed their worth, replacing surface mineworkers sent to fight on the Western Front and elsewhere.

Another important insight into the life of a Wigan pit brow woman working during the post-protest period was collected by local historian James Fairhurst interviewing Jenny Leyland in 2004. When Jenny was working at collieries from the 1930s onwards the numbers of brow women employed was related to the quality of the coal arriving on the surface. The older pits which were now working the poorer quality coal seams with high dirt content had a struggle to produce a saleable product and needed ever more women in the screening sheds. Also pits which did not have the capital to invest in coal washeries maintained their coal picking arrangements. On the other hand, new and large colliery sinkings from 1911 onwards more often than not included coal washeries, with few or no jobs for women.

Women's personal workplace conditions did not improve much either post-1911. Baths were a rarity in the Wigan area and only to be found at the larger companies' mines. They could afford to contribute substantially towards funds provided by the national Miners' Welfare Fund. Wigan Coal and Iron's Parsonage Colliery Leigh was an example, with baths provided from the early 1930s.

Selected extracts from Jenny Leyland's experiences follow:

A Pit Brow Girl's Story

Jim Fairhurst

The following was given to me by Jenny Leyland (nee Hooton) who is now well into her 80's, and living in Ashton-in-Makerfield. She started work on the pit brow of Long Lane Colliery, better known as 'Crowpit' in 1934 at the age of 14. Jenny said they started at 7a.m, had half an hour for breakfast and finished at 3:30. If it had been a bad day and there had been stoppages, they might have to work overtime, usually an hour. When she began work, two of the women, Mary Anne and Charlotte Davies were in charge, with a man in overall control. The coal, in 10 cwt tubs came up the pit, went over the check weigh machine to have its weight registered and then on to a tipper called a 'kecker' (or kicker). The tipper would deposit the coal on belts underneath.

Only Mary Anne and Charlotte had authority to stop the belts. The work was heavy but the knack was soon learned. Sometimes there would be lumps of coal with a band of dirt running through, and a chipper would be used to separate the coal from the dirt; or a very large lump of coal would come down the belts, which would have to be stopped until it was broken up. If the belts were stopped too long it might stop the winding coal, so it was essential they were kept going.

Compo Men

At the time working alongside the women were men who had been injured underground and were known as compo men, i.e. in receipt of compensation. 'We were sometimes grateful for a short stop so that we could have a rest', said Jenny. If there was a particularly bad tub of coal, they would have to climb on the belt and spade the discard off. One of the sisters, Mary Anne, would station herself at the bottom end of the belt where the coal went into the wagon. If any dirt had been let through, she would stop the belts and give everyone a mouthful. Occasionally the belts broke and a fitter would have to mend them, giving the girls another short rest.

There were few accidents in Jenny's time. 'I only remember one accident, and that was a compensation man had two fingers taken off. If there was an accident underground and the pit was stopped we were always found something else to do for there was always plenty of cleaning up.'

The Cabin

'The cabin was an old engine house heated by steam pipes with no stove then. A canteen was built but it was after I left. The tea in our cans was kept heated on the hot water pipes and was often stewed'. The girls would take it in turns to have their break, for the 'pit never stopped winding. When the canteen was built it had a stove and a rest room with easy chairs and a toilet. Before that, to answer a call of nature, we had to improvise and use a quiet spot out of the way with a plank and a bucket. We had nowhere to wash our hands afterwards'.

Their complexions would be guarded by make-up - Snowfire cream was put on first and the make-up on top of it. 'The dust would stick to it and come off when we washed ourselves. We put the make-up on just as we would if we were going out at night. Sometimes our hands hardened and cracks formed in the skin which would be painful in winter. Our hands never seemed clean, no matter how hard we scrubbed them, for the coal dust became ingrained.'

The Cattle Market

Entertainment was going to the pictures and dancing. On Saturdays and Sundays, the girls paraded along Lodge Lane and Gerard Street in Ashton, parades referred to as 'the cattle market'. The idea was to find a partner and raise a family. Some of the girls were really nice and were Sunday school teachers. 'We were never ashamed of our work because we were as smart as anybody when we were out'.

Even when we went to work we had clean clothes and we washed our hair every night. We would put brown paper under our shawls to keep the dust out but it got everywhere. As time went by, a spray was played on the dust which improved conditions a little. There were never any baths at the pit. I had to finish when I married in 1938, and even if a woman had to bring children up having lost her partner, she would not be re-employed. I remember one girl who had a son by a married man who was allowed to stay on. Her mother brought up the lad and the woman married someone else later. The custom then was that when a couple became serious about each other, marriage would be anticipated and the first child would soon follow the ceremony. There was no shame in it for that was the way of things.

Occasionally a girl would be let down as aforesaid and there would be shame attached to illegitimate births, the issue being known as 'chance childs'. But there were never what is known as 'one night stands' for the consequences would be too serious for the women.

A Happy Little Pit

'The Crow Pit was a happy little pit. The manager, Tom Jameson, was very friendly and I could talk to him just like I'm talking to you. Our reading matter would be "Woman's Weekly", "Red Star Weekly" and "Red Letter". A new shawl would be greeted with gasps of admiration and the owner would hold pride of place until the next one came along, for women competed with each other as to who had the nicest shawl'.

Asked about Mary Anne and Charlotte Jenny said she didn't think they had ever been courting but both were sociable enough. As far as she could remember, wages were about 2s. 6d. per day, 15s. a week. When they stopped overtime they were allowed to keep the money, which was usually spent on clothes. At the age of 18 they were paid the top rate of wages. Charlotte Davies was given the BEM on 1 January 1950, at the age of 63, for 50 years service to the coal industry.

Nationalisation of the coal industry in 1947 brought with it gradual increases in mechanisation of operations both below ground and on the surface. Coal grading, washing and screening could now be far more efficiently achieved without expensive manual labour. Brow women's jobs were protected as much as possible after 1947. Natural wastage and retirement meant their numbers gradually dwindled. By the mid-1950s only two Lancashire collieries employed pit brow women.

Right: *Not the sharpest of images recording the Taylor sisters at Abram Colliery around 1920, but worth including due to the classic pose the lady on the left adopts!*

Below: *Wigan Coal and Irons Co.'s extensive surface layout at Crawford Pit, Aspull, around 1908. Sunk in the 1840s the two shafts were 223 and 318 yards deep. The pit closed in 1928. The magnifying glass reveals brow women close to the coal wagons.*

Above: *An unusual arrangement of pit brow women at No. 5 Moor Pit, Aspull, around 1920. Fourth from the top on the back row is Alice Seddon who features in a different light in the next photograph. The pit shut in 1924, its sixty-eight surface workers moving to Meadow and Crawford pits.*

Left: *Alice Seddon, the Aspull pit brow girl featured in the previous photograph, is seen here proudly holding her daughter. Writers about pit women have often contrasted their lives at work and home.*

Opposite above: *Tyldesley coal pickers during the 1921 strike. Behind stands William Ramsden's Wellington Pit. Pit brow women are no doubt amongst the group.*

Opposite middle: *Coal pickers on the waste heaps of Tyldesley Coal Co.'s Cleworth Hall Colliery during the 1926 strike, including brow women. These enormous mountains of waste with their burning sulphurous hot spots gave the district a distinctive smell right up until the mid-1970s.*

Opposite below: *A happy bunch of brow women and girls around 1920 at Bryn Hall Colliery, locally known as Crippens Pit. Here we can see a wide range of outfits in use.*

View from the west of Parsonage Colliery, Leigh, in the early 1960s. Sunk from 1913 onwards and eventually reaching 1,012 yards the workings were for a while the deepest in Europe. The colliery for many years employed around fifty pit brow women.

Parsonage Colliery brow women having a 'brew' alongside the sidings in 1928; behind them is one of Wigan Coal & Iron Co.'s distinctive 10-ton wagons. Note the typical iron-shod clogs with pointed tips.

Forty-five pit brow women feature in this shot taken at Parsonage Colliery, Leigh, in 1940. As this was wartime, they wore lightweight compressed card safety helmets. The wagon behind them to the right had come all the way from South Wales Amalgamated Anthracite Collieries.

Two brow women can be seen in the canteen at Parsonage Colliery in 1964 amongst men fresh out of the pit with black faces. The price board to the left shows that tea, milk, coffee, biscuits, cakes and pasties were all priced the same at 1s (5p today).

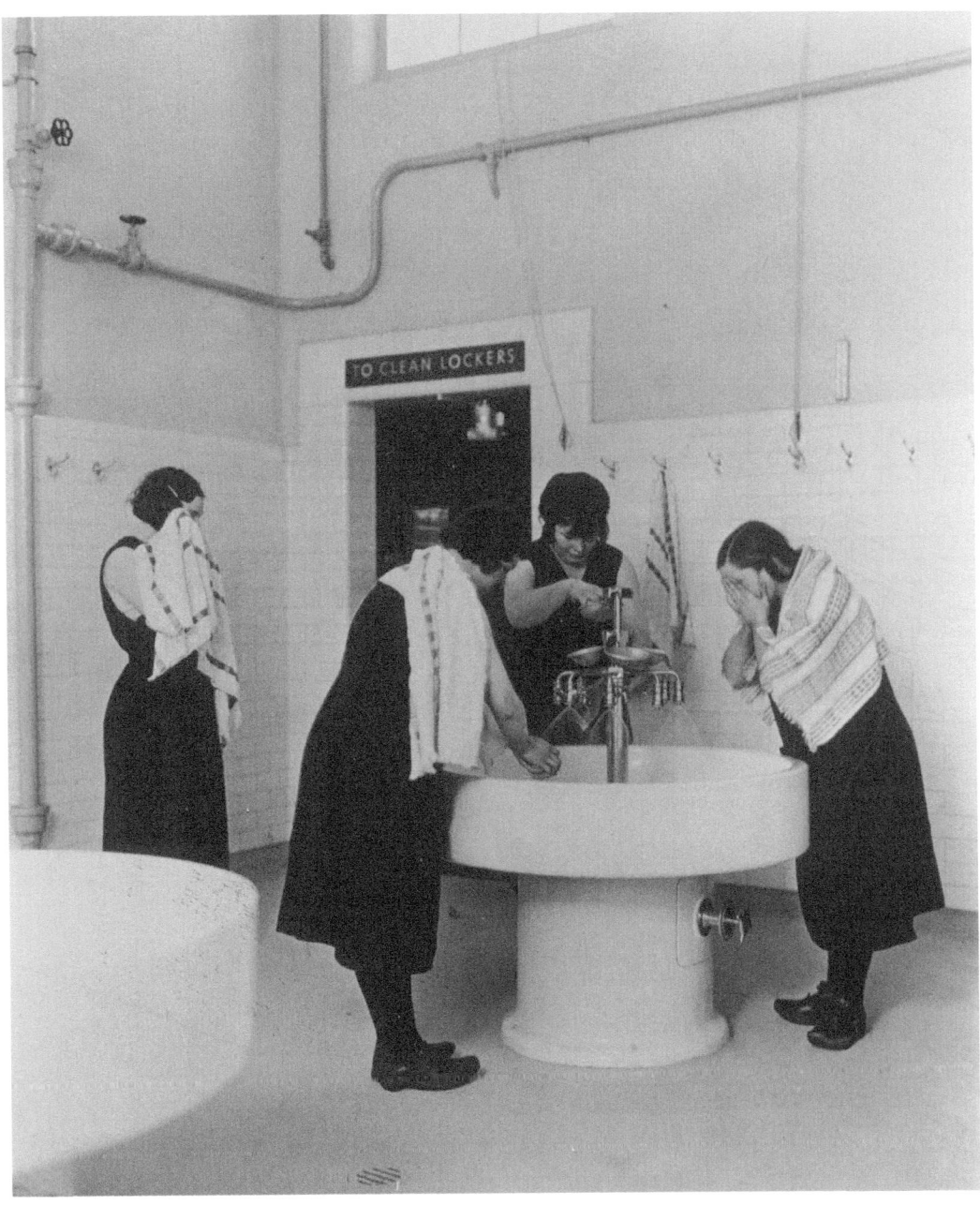

Above: *In the women's baths at Chisnall Hall Colliery shortly after it opened in 1934. Not all colliery companies provided separate facilities; Wigan Coal Corporation could do so, being such a massive concern.*

Opposite above: *Ince Moss Colliery, Wigan; brow women Phyllis Sherrington, aged twenty-two, and her friend Cilla Grimes, aged twenty-four, pose on the pit bank in the late 1930s. The pit shut in 1962.*

Opposite below: *Chisnall Hall Colliery, Coppull, around the late 1930s, a Wigan Coal & Iron Co. colliery. Brow women had their own baths here after 1934. The pit shut in March 1967.*

The brow women's rest room at Chisnall Hall Colliery, opened in January 1935. Furnished in classic Art Deco fashion, the room includes two of the famous Mies Van Der Rohe Brno chairs designed at the Bauhaus design centre in Germany. These are now worth thousands of pounds!

Nook Colliery, pit nos 4, 3, 1 and 2, Tyldesley, in the 1930s when part of the giant Manchester Collieries concern. No.4 pit reached the Arley seam at 944 yards where temperatures often reached 100F.

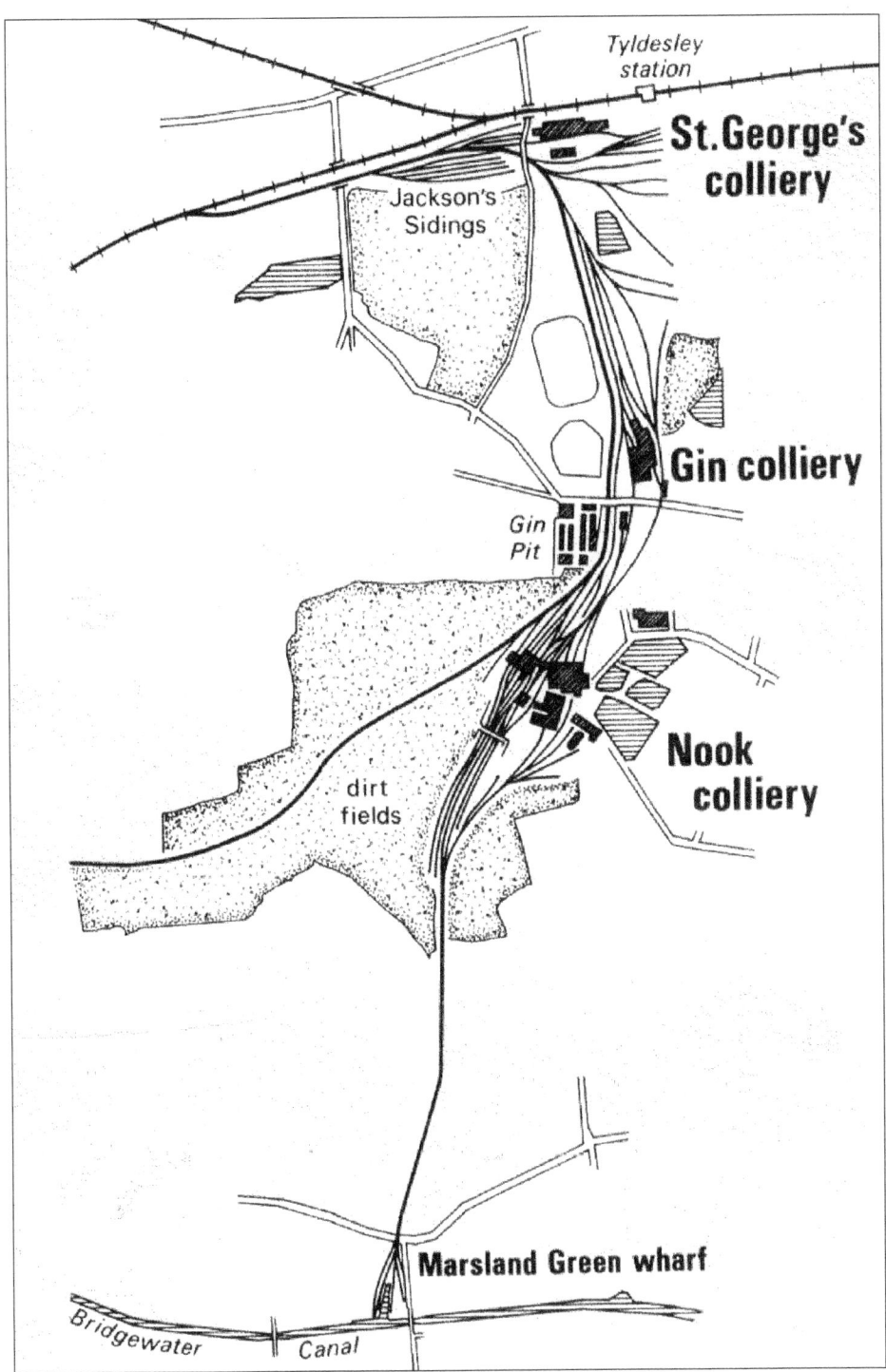

Collieries around Tyldesley in the early 1950s, which employed women on the pit brow. Many would live in the isolated pit village of Gin Pit which still survives today, even though the collieries have long gone.

A happy surface group at Nook Colliery, Tyldesley, in the 1920s, the women wearing a variety of costumes.

Nook Colliery pit brow women and two young surface lads around the late 1920s to early 1930s. The lads are sat on some old iron screening plates. Used in the screen sheds, these allowed set sizes of coal through to the Manchester Collieries wagons you can see in the photograph. Young lads often passed through 'initiation ceremonies' involving grease but that is another story!

Pit brow women at Wigan Junction Colliery near Abram help load coal from the pit's stock heaps. The tubs were hauled up an incline by the colliery winding engineman and tipped into main-line wagons. It was the severe winter of 1940 when snow lay on the ground for over a month with drifts up to 6ft deep. As the colliery was out of action the coal tips acted as a reserve during this period when demand was high and production nationally low. From the left: Margaret Jane, Annie Pugh, Mary Wall, Edith Webb, Daisy Nixon and Lena Adamson.

With the war two months old, the new pithead baths for women opens at Wigan Junction Colliery on 18 November 1939. Mrs Onions, wife of the well-known miners' agent for the area, stands to the left of the door.

Left: *Former pit brow woman Mary Redford (née Wall) photographed as this book was finished in May 2006. She can be seen in the previous winter scene at Wigan Junction. Born in 1921 during the miners' lockout, she worked a six-day week during the Second World War, working Monday to Friday 7.00 a.m. to 2.30 p.m. and Saturday 6.00 a.m. to 12.30 p.m. She recalls wearing a black outfit with thick black stockings overlapped by her huge black elasticated knickers! During the war, local pit brow women were asked to 'Dig For Victory' in their spare time but the land they were given was useless, with only radishes managing to grow. She recalls not being allowed to wear gloves on the brow during the 1940 winter, the unfriendly foreman (who always called the women 'Yo', rather than their names) saying she could not do her job as efficiently wearing them.*

Below: *Brow women and male surface workers inside 'C Shelter' at Wigan Junction Colliery during the Second World War.*

Pit brow women at Wigan Junction Colliery (seen behind) in July 1941 after completing a first-aid course. From the left: Mary Wall, Margaret Lomax, Doris Mills, Margaret Hill, Mrs Teresa Redford (née Wall) and Mary Bramhall.

A poor image but an unusual one showing the Christmas party for surface workers at Chanters Colliery, Atherton, around the early 1940s. Atherton Collieries distributed either a food hamper from Rushtons of Wigan or a large turkey to its workforce. Chanters took on the local nickname of Turkey Pit.

Men working alongside pit brow women in the screen sheds at Gibfield Colliery, Atherton, in the late 1930s. Any dirt, shale or stone passing by was thrown into the central shaker conveyor. It then headed off to be tipped on the waste heaps near Schofield Lane.

Screen women at Chanters Colliery, Atherton, in the late 1930s, a photograph taken for a feature on the industry. The women are chipping off shale or stone from the large cobs of coal passing by.

A jolly bunch of brow women pose for the cameraman in 1939 at Long Lane Colliery (Crow Pit), Bryn, near Ashton.

Surface workers pose in the early 1950s at Mains Colliery near Bamfurlong. By this time women just wore whatever old clothes they liked as compared to the old costumes.

A fascinating scene in the stockyard at Garswood Hall Colliery around 1940. The pit headgear pulley wheel or sheave standing alongside would be a spare one in case of damage to those in use. The photograph shows how large these pulleys were close up.

A smiling group at Garswood Hall Colliery Top Pit (No. 6 Pit) around 1951 to 1952. The shaft behind was 700 yards deep. The colliery closed in 1958, although carried on washing coal on site until 1962. Three Sisters Country Park occupies the site today.

Garswood Hall Colliery again in 1951/52 showing how dirty men could get purely working on the surface.

Victoria Colliery, Standish, around the early 1940s. Sunk by Wigan Coal & Iron Co. in 1900, it was closed in 1959. Women worked here on the screens until at least 1953.

Freda Farrimond visited her great aunt, a pit brow lass at Victoria Colliery, Standish, in the early 1950s. Her reminiscences were published in Wigan *Heritage Services Past Forward Magazine* No.11 in autumn 1995. They give us an insight into the final phase of the employment of pit brow women in the Wigan area. Selected extracts follow:

During my school summer holidays in the early 1950's I often visited Victoria Colliery with my father and uncle who were coal merchants. I was given the opportunity to visit my great aunt at work. Both my grandmothers worked at Broomfield and Prospect Pits (Standish). The pit brow lasses who worked at Victoria Colliery were the last to be employed at a Standish Colliery.

My great aunt worked for 47 years on the 'shakers' (oscillating metal screens grading the coal). Having started work in 1906 at 13 years of age at Prospect Pit then transferring to Victoria Colliery in 1908, she remained there until her retirement in 1953 at the age of 60 years. Life on the pit brow was arduous, but nevertheless the women always seemed cheerful. In winter the wind blew across the elevated pit brow making it extremely cold and draughty. The women had to work with their coats on, and wore gloves without fingers on their icy hands.

In summer, by contrast, the pit brow was very hot and dusty, the women wore long sleeved overalls to protect their arms from the dust, which could cause dermatitis. They also wore dust caps to cover their hair, black stockings and clogs. Due to the lack of adequate washing facilities on the pit brow the women could only wash their hands before lunch. Soap and towels were however issued to all colliery workers annually. My great aunt carried her soap and towel and lunch to work in a small case together with her cup and three small tins containing tea, sugar and Nestles milk.

At the end of the working day all the pit brow lasses would walk up Lurdin Lane together to catch the bus into Standish. My great aunt always called at our house on her way home, as my grandmother (her sister) had a cup of tea and a snack waiting for her. Both of them would then listen to Mrs Dale's Diary on the radio, but my great aunt often fell asleep in the chair before the episode had ended. She was very tired and deserved a well-earned rest after a hard day at work on the pit brow.

Freda's reminiscences are interesting as her great aunt worked at Wigan Coal & Iron Co.'s Victoria Colliery for nearly its whole life. It was sunk in 1900 and closed in 1959.

Ethel Hampson worked at Nook and Gin pits at Tyldesley in the inter war years. Aged seventy-seven in 1984, she remembered that not only coal dust made the job a dirty one. Collieries dispensed with the idea of toilets below ground after occasional trials had proved unsuccessful. Ethel comments, 'So the shit would come up with the coal sometimes. I could smell it coming, so I'd shout: Here come the ashes o't roses, watch out!' She also remembers throwing 'lumps of coal as big as a television. You had to have muscles like Primo Carnera'.

Mary Appleton, born in the 1930s, reminds us, 'of course you got dirty and there wasn't a house with a bath except the boss's. I used a spade bigger than myself to get the dirt into the wagons'.

Etty Ode of Ince remembered in 1984 that, 'I used to swing a 7lb hammer to smash the kennel' (cannel coal, very hard and splintery).

Margaret Broxson recalls that trousers made a re-appearance after the Second World War when she started work on the brow. The younger women, like herself, updated the image of brow women, 'The old women wore old-fashioned skirts and shawls. We wore coloured turbans.

You brassy little buggers they would say. And we got rid of the black stockings and wore pants instead. There was uproar. Oh we were brassy!'

Jim Fletcher, manager of Howe Bridge Colliery, Atherton, during the Second World War had a number of women and girls at work. 'I had 36 picking coal and we always had to allow two spare… not for absentees but because they always went to the lavatory in pairs throughout the day. I never knew whether they worked on a rota or not'.

By 1953 pit brow women in the Lancashire coal field numbered just less than 1,000. Around this time a deal was struck between the National Coal Board and the National Union of Mineworkers which meant that the pit brow women's days were numbered. Bernard Donaghy, a union official in the Lancashire Area NUM at the time and later to become area president, remembers that the decision to give elderly or injured miners priority as regards surface work was taken at one of the Lancashire Conference meetings, with representatives from all the collieries present. He adds that in all honesty the women already doing the job were often stronger than the men who replaced them!

The last ever Lancashire pit brow woman coincidentally worked at a Wigan area colliery, Golborne, until 1966 and the last pit brow woman employed nationally ended her working days at Whitehaven in 1972. Further afield, removal of women and girls from mines was only finally achieved in Belgium, for instance, by 1911. Today women and girls can still be found in mines in America (3,300 in 2002), India, Cambodia and Columbia and many other remote areas worldwide. Women only work on the surface today at Chinese mines.

Above: *Standish Hall drift mine around 1952. Women seen here are picking out dirty coal from the conveyor belt.*

Right: *Standish Hall drift mine around 1952. Surface workers included Ann Leach, far right.*

Opposite above: *Inside the screen sheds at Victoria Colliery, Standish, in 1908. Look out for the reminiscences by Freda Farrimond recalling her great aunt working on the brow here, featured in this book. She is probably in the photograph.*

Opposite below: *Standish Hall drift mine, Elnup Wood, around 1950. This short-lived mine accessed a small area of coal reserves opening in late 1948. A small number of women worked here on the surface until it closed in 1961.*

Left: *Screen women leave the pithead after their shift in the early 1950s. This was probably Giants Hall Colliery, Standish. A smiling Sally Gittoes is in the foreground.*

Below: *Pit brow women at Astley Green Colliery pose for a group photograph on 18 July 1952. The colliery closed in 1970 and is now a mining museum, well worth a visit.*

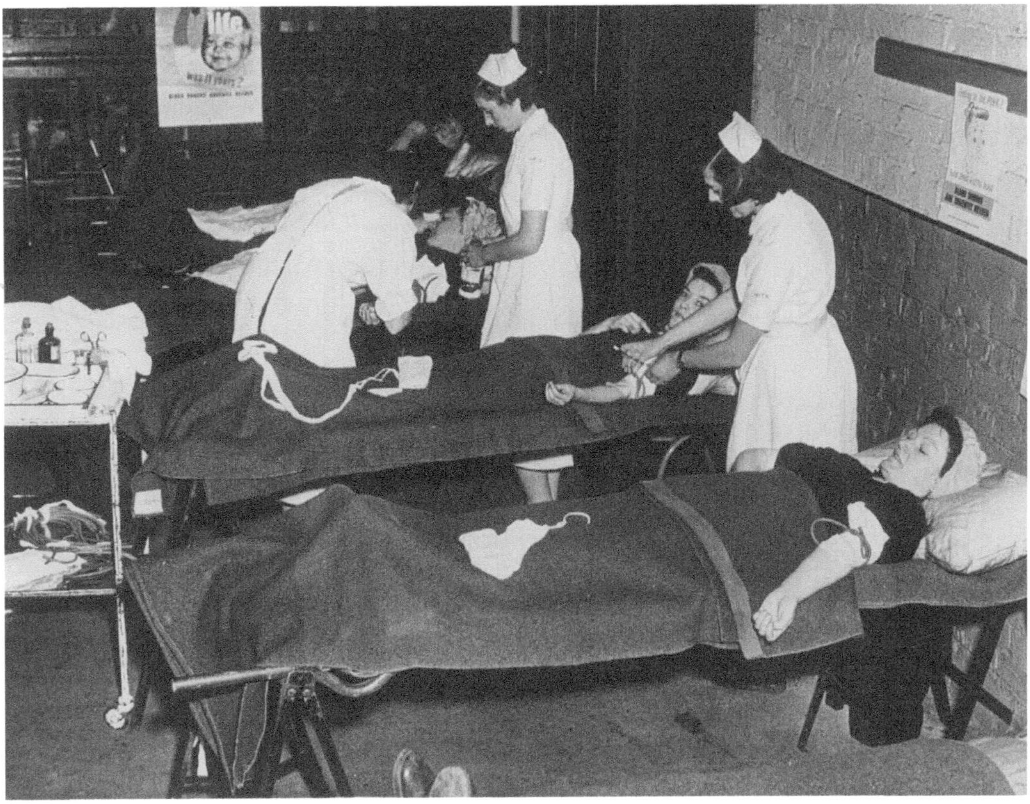

Pit brow women giving blood at Astley Green Colliery in May 1953.

Pit brow women's farewell party photograph taken at Astley Green Colliery in August 1955.

We should not forget the female medical attendants and nurses who worked at many Wigan area collieries. They saw severe and often fatal injuries virtually on a daily basis and were the only women allowed to work below ground by law. Here three have attended a course at Boothstown Mines Rescue Station, Ellenbrook Road, Worsley, in 1954.

Appendices

Appendix 1

Serious and fatal accidents suffered by Wigan area pit brow women and women straying onto the colliery surface as reported by the Mines Inspectorate 1854 to 1918

[Note: Apart from the first five entries which are incomplete these are arranged in the following order: Name of woman / girl. Date of accident (y/m/d). Age. Job. Colliery. Colliery Owner. Colliery location. Type of accident]

Farrimond, Elizabeth 1854/10/3 16   Springs Colliery   Thomas Pearson   Wigan
Run over by railway wagon near pit.

Marsh, Mary 1854/11/25   Orrell Colliery   W.H. Brancker and Co.   Wigan
Injured on surface railway.

Moss, Ann 1855/1/22   Ince Hall Coal and Cannel Works   Ince Hall Coal and Cannel Co.   Wigan
Crushed by wagon at Arley pit.

Golding, Mary 1856/8/23   Pemberton Colliery   Jonathan Blundell and Son   Wigan
Being run over by a railway truck near the pit bank.

Pennington, Mary 1861/10/22   Ince Hall Colliery   Ince Hall Coal Co.   Wigan
Crushed between flywheel of engine and foundation.

Barker, Margaret 1865/2/23 17 Coal setter   Springs Colliery   Pearson and Knowles   Wigan
Crushed between coal trucks above ground.

Broadhurst, Sarah 1865/9/9 29 Coal dresser   Kirkless Hall Colliery   Kirkless Hall Coal and Iron Co.   Wigan
From injuries by coal trucks on surface railway.

Tabes, Hannah 1867/11/6 32 Labourer   California Pit   Wigan Coal and Iron Co.   Wigan
Crushed by coal trucks on surface.

Hackett, Alice 1871/2/3 20 Coal cleaner   Hindley Colliery   Wigan Iron and Coal Co.   Wigan
From a wound in the leg by a fork above ground.

Walkden, Ellen 1874/8/1 18 Screener   Ladies Lane Colliery   Wigan Coal and Iron Co. Ltd   Wigan
Run over by wagons on surface.

Name	Date	Age	Occupation	Colliery	Company	Location
Walls, Ann	1874/8/13	55	Screener	Bradshaw Colliery	Wigan Coal and Iron Co.	Wigan

Crushed between wagon and wall.

Catteral, Mary	1880/3/14	20	Screener	Brynn Moss Colliery	Smethurst Hoyle and Grime	Wigan

Found drowned in the canal at colliery. There was no evidence to show how the accident occurred.

King, Matilda	1881/5/9	16	Screener	Pemberton Colliery	J. Blundell and Son	Wigan

Inquest attended. The deceased and another girl were engaged cleaning up the screens and she went to push a wagon under the screen to put the rubbish in when some others were pushed against it by two men who were not aware of her presence and she was crushed between the buffers. There was no necessity for her to touch the wagon and she had been specially warned a few days previously on no account to interfere with any of the wagons.

Jones, Priscilla	1881/5/12	4	Child	Robin Hill Colliery		Wigan

Killed by a tub which rolled down dirt heap where she had been placed by an aunt. Not comprised.

Taylor, Ellen	1882/7/11	15	Screener	Bamfurlong Colliery	Cross Tetley and Co. Ltd	Wigan

Inspection made. Deceased fell into a tank of hot water when she went to fill her can and was scalded to death. The tank was 2 feet square and 8 feet high and appeared to be perfectly safe.

Eastham, Jane	1884/2/21	20	Screener	Hindley Green Colliery	J. Scowcroft and Co.	Wigan

Inquest attended. The deceased and some other girls were passing between the buffers of some wagons which were standing a few feet apart when some other wagons which were being lowered ran too fast owing to the slippery state of the rails and bumping up against the standing wagons knocked them over the scotches. The manager promised to make such arrangements that it would not be necessary for the girls to cross the line in future.

Hilton, Theresa	1884/5/15	16	Coal picker	Hic-Bibi Colliery	Ellerbeck Collieries Ltd	Chorley

A prop which she put in front of her wagon was thrown against her.

Kay, Mary	1885/6/16	24	Coal picker	Hindley Green Colliery	John Scowcroft and Co.	Wigan

Deceased was at work in a wagon when she was struck by a tub which fell down the screen having by some means slipped out of the tippler.

Turton, Faith	1886/1/15	16	Wagon trimmer	Hindley Deep Pit	Wigan Coal and Iron Co. Ltd	Hindley

Inspection made and inquest attended. The wagon lowerer lowered some wagons without looking to see that the road in front was clear and his wagons struck against the one the woman was on and knocked her off under the wheels.

Platt, Margaret	1886/7/29	21	Screener	Fir Tree House Colliery	Crompton and Shawcross Ltd	Wigan

Inspection made. Deceased was picking out stones on the top of a loaded wagon and was told to get off as the wagon had to be moved. She neglected to do so and when the other wagons bumped up against it she was knocked off and fell under the wheels.

Leyland, Elizabeth	1886/12/2	25	Screener	Alexandra Colliery	Rose Bridge & Douglas Bank Collieries Co. Ltd		Wigan

Deceased was supposed to have strained herself but went on working till the 4th inst. when she was taken ill and died on the 14th inst. from typhus fever.

Davies, Lucy Ann	1888/5/16	30	Number taker	Brinsop Hall No.6 Siding	Brinsop Hall Colliery Co. Ltd	Blackrod

Four ribs between the buffers of two railway wagons. She was standing against the buffer wagon with her back to one which the wagon lowerer was letting down and he not did not call out.

Dickinson, Ellen	1889/2/27	52	Not employed	Scot Lane Colliery	Scot Lane Colliery Co.	Blackrod

Killed by the colliery locomotive engine as she was making a short cut across the colliery and No.5 Pit the night being dark. Non-Comprised. Inspection and inquest attended.

Cartwright, Martha	1889/5/6	17	Coal picker	Atherton	Fletcher Burrows and Co.	Atherton

Fingers crushed. Drum which carries the conveying rope bands at the coal cleaner. Inspection and inquiry.

Greenall, Moira	1890/1/16	24	Screener	Orrell Colliery	Orrell Coal and Cannel Co. Ltd	Wigan

Inquest attended. She was passing across the line to her work ands was going between two wagons when the engine suddenly backed and she was crushed to death.

Gore, Margaret	1891/5/4	14	Coal cleaner	Blainscough Hall Colliery Siding	Blainscough Hall Colliery Co.	Coppull

Fatally injured. Arm ran over by a railway wagon that was moved when she was getting down. Inspection and inquiry. Died 8th at Wigan Infirmary. Arm amputated 7th. Attention at the inquest was called to time being scarcely allowed for her to get clear and to the danger of delay in amputating in such cases. Inquest attended.

Fields, Mary	1891/6/3	25	Wagon trimmer	Low Hall Colliery	Moss Hall Coal Co. Ltd	Wigan

Inspection and inquiry made. This woman had to trim wagons and with the intention of getting up on one on the brake side she went between the buffers of a standing wagon and one being lowered by hand and was crushed. She could have gone behind the wagon quite as readily.

Prescot, Ann	1894/7/5	15	Screener	Abram Colliery	Abram Coal Co. Ltd	Bickershaw, Leigh

Inspection made. While in the act of coming down the steps from the screen to which there was a hand rail attached she accidentally fell and received injuries the following day.

Makinson, Ellen	1894/10/17	18	Tub cleaner	Crawford Pit	Wigan Coal and Iron Co Ltd	Aspull

Inspection made and inquest attended. There were some tubs standing at the top of the wooden inclined stage and the deceased appears to have pushed one over and got in front to try to stop it when it overpowered her and causing such injuries that he died on the 31st instant.

Oswald, Ellen	1898/7/26	16	Screen hand	Garswood Hall Colliery	Garswood Hall Collieries Co Ltd	Ashton-in-Makerfield

Inspection made and inquest attended. Three girls sat down on a fence rail the edge of the pit black and it broke suddenly being partly rotten with age causing them to fall a distance 25 feet to the ground. The bad state of the rail was partly hidden being bad where it was socketed into upright. She succumbed to the injuries received on the 31st instant.

Mason, Elizabeth A.	1901/12/28	13	Pit brow girl	Middle Place Pit	Garswood Coal and Iron Co	Ashton-in-Makerfield

Inspection made and inquest attended. Deceased had been paid her wages and had to cross the railway lines near the screens to go home. She was knocked down and run over by the two empty wagons. The loco was pushing the two wagons and one of the brakesmen ought to have ridden on the leading wagon but he neglected to do so hence the accident which otherwise would have been prevented. She died from her injuries received at Wigan Infirmary Jan. 25th.

Prescot, Mary	1902/11/7	18	Screen hand	Wigan Junction	Wigan Junction Colliery Co.	Wigan

Inspection made and inquest attended. Deceased attempted to cross some bevelled pulleys where a face rail had been taken off near the screens and her dress got caught in the wheels and she got her leg severely injured. She expired at the Wigan Infirmary 5th 1903 after undergoing an operation. Manager prosecuted and convicted for non-fencing.

Powell, Margaret	1910/6/4	23	Pit brow girl	Low Hall Colliery	Moss Hall Coal Co Ltd	Wigan

Whilst on her way to work and whilst crossing Moss Hall Co.'s railway, Powel was knocked down by a wagon in course of shunting and killed. She is reported to have been employed at Wigan Junction Colliery.

Brown, Ethel	1916/6/29	19	Picker on belts	Nook Colliery No.3 Pit	Astley and Tyldesley Collieries Co Ltd	Tyldesley

Crushed by tub, fractured arm.

Heyes, Mary	1916/10/30	20	Coal tippler	Low Hall Colliery	Moss Hall Coal Co. Ltd	Wigan

Crushed between tipple and tubs, severe contusions.

Warburton, Maggie	1917/8/24	30	Labourer	Bamfurlong Colliery	Cross Tetley and Co. Ltd	Ashton-in-Makerfield

Crushed between barrel and tub, crushed hand.

Acton, Edna	1918/9/24	17	Coal picker	Tyldesley Colliery	Astley and Tyldesley Collieries Co Ltd	Tyldesley

Fell in wagon, injured face.

Appendix 2

A Veteran Pit Brow Lassie
Retires after sixty-two years of work

Wigan Observer, 28 August 1920:

Miss Ellen Wilkinson of 4 Landgate Lane, Bryn Ashton-In-Makerfield, who is now seventy three years of age has retired from the position of pit brow worker, after following the employment over sixty two years. She was born on 27th April 1847 at Chapel Lane Wigan and lived there for about six years, when she came with her parents to Ashton-In-Makerfield, where she has lived ever since. Her father was browman at what was known as the Old Five Feet Top Colliery, North Ashton, of the Garswood Coal and Iron Co. and before his daughter Ellen was eleven years of age she commenced work at the colliery. Her father used to take her with him about half past four in the morning and bring her back at night about 5.30 pm. After she had got well used to her work she was paid the handsome sum of 8d per day and in those early days she thought herself exceedingly well off. Her parents died over thirty years ago, leaving two unmarried sisters and another the wife of the late Councillor Elias Woods, brother to the late Mr Sam Woods M.P for Wigan. All three sisters are still living in Ashton-In-Makerfield. There have been several women pit brow workers who have put in a long number of years at their work in the Wigan district but the case of Miss Wilkinson probably constitutes a record.

Appendix 3

The collier lass

Lancashire song of unknown origin, c.1842:

My names Polly Parker, I come o'er from Worsley
My father and mother work in the coal mine
Our family's large, we have got seven children,
So I am obliged to work in the same mine.
As this is my fortune, I know you'll feel sorry
That in such employment my days I shall pass
I keep up my spirits, I sing and look merry
Although I am but a poor collier lass.

By the greatest of dangers each day I'm surrounded
I hang in the air by a rope or a chain.
The mine may fall in, I may be killed or wounded,
May perish by damp or the fire of the train.
And what would you do if it weren't for our labour?
In wretched starvation your days you would pass,
While we could provide you with life's greatest blessing,
Then do not despise the poor collier lass.

Now all the day long you may say we are buried,
Deprived of the light and the warmth of the sun;
And often at nights from our bed we are hurried:
The water is in and then barefoot we run.
And though we go ragged and black are our faces,
As kind and as free as the best we'll be found;
And our hearts are as white as your lords, in fine places,
Although we're poor colliers that work underground.

I'm growing up fast, now, one way or another;
There's a young collier lad strangely runs in my mind.
In spite of the talking of father and mother,
I think I should marry if he was inclined.
But should he prove surly and would not befriend me,
Another and better chance may come to pass.
My friends here, I know, to him will recommend me,
And I'll be no longer a collier lass.

Appendix 4

Breaking a man's arm with a poker

Wigan Observer, Friday 20 April 1860:

Ellen Sharples, a middle aged woman, who is employed on a coal pit brow, was charged with assaulting James Whitter, of Standish, collier. - Mr. Mayhew appeared for the defendant. - The complainant said he was going home on Saturday night last, and on passing the defendant's house at Douglas Bank she called him in, to ask him some questions in reference to her son. He told her that her son was not worth talking about, whereupon she got very violent, and after some words had been exchanged she seized the poker and attempted to strike him on the head. He, however, put up his arm to ward off the blow, and caught the weapon upon the joint. He then received two other blows, and was disabled, medical attendance having to be called in. - The defence was that the complainant was the aggressor, the defendant's version of the affair being that the complainant, on the night in question, walked into her house unasked, in a state of intoxication, and wanted some tobacco given him. This was not supplied, and he pulled out three halfpence to pay for some, but he was told he was not wanted, and had better go out of the house. He accordingly left, but returned directly afterwards, and commenced to assail defendant with all sorts of foul names, becoming exceedingly abusive. An attempt was made to put him out of the house, and the defendant admitted that in her anger she took up the poker and struck the complainant. - The bench said an assault had been committed, but as there were extenuating circumstances in the case they should only inflict a fine of 10s and costs.

Appendix 5

A Pit Brow Protest?

H.D. Rawnsley
Wigan Observer, 3 April 1886

> Is honour dead, is justice out of date,
> That thus your law-amenders tyrannise?
> Shall envious greed attired in angel guise
> Of morals, decency and health dictate?
> No, from the dark pit's brow where quaintly dressed
> But not unsexed and free from taint of ill
> There toil the sisters of the whirring loom.
> Let the brave lads of Lancashire attest
> Though hands are hard, yet hearts are gentle still
> And maiden blushes mount where roses bloom.

Appendix 6

Boompin' Nell

Poem by A.J. Munby of 1887 (Victorian diarist, poet and barrister, 1828–1910)

'Couldn't I work theer ageean Sir? I'd go as snug as I can;
I'd cut my hair quite short, I would, and I'd dress mysen like a man.
Why, I've getten my breeches here, an my owd top coat an all,
An I lay they suits me better, nor a Sunday bonnet and shawl;
For my showthers is rare and broad, you see, an I never had much of a figure,
An my hands is as hard as nails, look, an as big as a man's, or bigger;
An' what is a woman's voice, when you shouts, but never speaks?
An' my face 'ud be thick in coal dust, so they'd never notice my cheeks,
An' the best of it is, I are tall, so they wouldna make much out o' me,
If I once goes down like a man, Sir, why a man I can easy be.
My mates 'ud know me, it's true; but they wouldna tell, not they!
So, if you'll only let me, why I'll do it, straight away'.

Appendix 7

On the brow

Poem by A.J. Munby 1892:

Give me thy hand, high-couraged lass but coy-
The hand that now so deftly grasps thy spade:
Perchance 'twill tell me how sweet a maid
Can find her calling in such rude employ.

Surely thy life, thy work, should be all joy,
And fair as thou art; yet thou hast essay'd
A strong man's labour! Art thou not afraid
Such tasks as these thy beauty will destroy?

'Nay Sir,' she said, 'I care na if they do!
My hands is what I live by, not my face.
I are no lady, nor no man like you,
As maybe thinks it inna woman's place
To choose her work. Take hold then, an you'll see,
What sort o' hand bets suits a wench like me!

Appendix 8

A Pit Brow Wench for Me

Author unknown, *c.*1890

I am an Aspull collier, I like a bit of fun
To have a go at football or in the sports to run
So goodbye old companions, adieu to jollity,
For I have found a sweetheart, and she's all the world to me.

Could you but see my Nancy, among the tubs of coal,
In tucked up skirt and breeches, she looks exceedingly droll,
Her face besmear'd with coal dust, as black as black can be,
She is a pit brow lassie, but she's all the world to me.

Appendix 9

'Delicate' Frances celebrates after time on pit brow

Lancashire Evening Telegraph, Thursday 11 June 1998:

A former Pretoria Pit (Atherton) brow lass, put into the job because she was 'delicate', celebrates her 100th birthday on Sunday. Lively Mrs Frances Butler who still does her own shopping each Monday, has loved all her life in Atherton. She married miner James, who passed away 27 years ago, and has two surviving children Elsie, 78, and Jean, 63. 'She started work at the Pretoria as a pit brow lass because she was considered by her family to be delicate and needed the fresh air. She is anything but fragile now'.

No. 3 of a series of postcards entitled 'Pit Brow Girls' dating from about 1905. The series was taken at Wigan area collieries, this one at Moss Hall Colliery, Arley Pit. In the background is the surface winch for raising and lowering tubs.

Appendix 10

Carnival float plea

Lancashire Evening Telegraph, Thursday 3 July 2003:

Shawls, flat caps and clogs are desperately needed by Gin Pit (Tyldesley) villagers who aim to enter a float in Leigh Carnival with a coal mining and pit brow lass theme.

Bibliography and Sources

Wigan Leisure and Culture Trust
Leigh Library Local Studies Dept
Wigan Leisure and Culture Trust
Wigan Archives
Leigh Town Hall
The Wigan Observer
The Leigh Journal
The Colliery Manager
Pictorial World
The Graphic
The Illustrated London News
The Black and White Budget, 1901
Mrs P. Holden, *The True Story of a Lancashire Pit Brow Lass*, 1947
Bergarbeiterinnen, Christine Vanja, Der Anschnitt, 39, 1987
'Delivering Coal Mines: Advertising Photographic Postcards of the Coal Mining Industry in England', Yukiko Inui, *Iconics* Vol.7, The Japan Society of Image Arts and Sciences 2004
Munby, Man of Two Worlds, Derek Hudson, Abacus 1974
Colliery Legislation and Its Consequences: 1842 and The Women Miners of Lancashire, Angela John, Rylands Library, Manchester 1978
The Coal Pits of Chowbent, Ken Wood, K. Wood 1984
Victorian Working Women, Michael Hiley, Gordon Fraser Gallery 1979
The Lancashire and Cheshire Miners, R. Challinor, F. Graham, Newcastle 1972
Living and Working In Wigan, J. Hannavy and C. Ryan, Smiths Books, Wigan 1986
Lancashire Magazine 1988
The Observer 1979, 1984
Mines and Miners of South Lancashire 1870-1950, Donald Anderson and J. Lane, D. Anderson 1980
Our Past Our Struggle, Rochdale Art Gallery touring exhibition catalogue, Sarah Edge, Rochdale Art Gallery 1986
Coal mining in the Eighteenth and Nineteenth Centuries, B. Lewis, Longmans, London 1971
The Miners, A History of the Miners Federation of Great Britain, 1889-1910, R. Arnot, Allen & Unwin, London 1949
The Miners, The History of the Miners Federation of Great Britain, from 1910 onwards, R. Arnot, Allen & Unwin, London 1953
The Miners, In Crisis and War from 1930 onwards, R. Arnot, Allen & Unwin Ltd
Howarth, Ken, *Dark Days*, Greater Manchester County Council, 1978
The Coal Industry of the Eighteenth Century, T.S. Ashton and Joseph Sykes, Manchester University Press, 1964
Coal Mining, Dr A.R. Griffin, Longman, 1971
The Great Northern Coalfield 1700-1900, Durham County Local History Society, 1966
Coal Mining and the Coal Miner, H.F. Bullman, Methuen & Co., London 1920
The History of the British Coal Mining Industry. Vol.1 Before 1700, Dr J. Hatcher, OUP
The History of the British Coal Mining Industry. Vol.2 1700-1830, M.W. Flinn, OUP
The History of the British Coal Mining Industry. Vol.3 1830-1913, R. Church, OUP
The History of the British Coal Mining Industry. Vol.4 1913-1946, Prof. B. Supple, OUP
The History of the British Coal Mining Industry. Vol.5 1946-1982, W. Ashworth, OUP

Records of Mining in Winstanley and Orrell, Near Wigan, J. Bankes, Lancashire and Cheshire Antiquarian Society, Vol. 54, 1939

Lancashire and Cheshire Miners Federation minutes

Other titles published by The History Press

Women at Work on London's Transport 1905–1978
ANNA ROTONDARO

This evocative collection of images charts the history of women at work on London's transport, from a typist employed by the District Railway in 1905, to the first women bus and tube drivers in the mid-1970s.

0 7524 3265 6

Wigan History and Guide
JOHN HANNAVY

This book gives the history & guide of Wigan: one of the oldest chartered boroughs in the north of England. The book includes two walking tours of the town which can be used independantly of the main text.

0 7524 3099 8

Collieries of North East Lancashire
JACK NADIN

Jack Nadin's compilation of images from the collieries of North East Lancashire, covering Hapton Valley, Bank Hall, Copy, Deerplay, Old Meadows, Salterford, Gambleside, Hoddlesden, Huncoat, Reedley, the Fence Drift mines, as well as some of the private pits around Burnley.

0 7524 2803 9

If you are interested in purchasing other books published by The History Press, or in case you have difficulty finding any of our books in your local bookshop, you can also place orders directly through our website

www.thehistorypress.co.uk